"What is interesting about this book is its breadth. Even though its title suggests only manufacturing applications, it demonstrates how to use the Kano model and a wide variety of quality tools and methods to build a customer-driven organization of any kind. It takes a wide view of application rather than being in depth. It thereby sets the broader context for really understanding how to achieve customer delight rather than getting bogged down in the tools. After all, what really counts is how the principles, tools, and methods are integrated together to achieve excellent products or services—that is what the customer seeks. That is what this book achieves."

—Frank Murdock, MSIE, MS Applied Mathematics; 2014 Chair, ASQ Lean Enterprise Division; Principal Consultant, FKM Consulting

"This book shows how the Kano model, Voice of the Customer, and Quality Function Deployment bring successful products into the marketplace for both service and manufacturing companies."

—James Bossert, PhD, CLSSMBB; 2014 Editor *ASQ Six Sigma Forum Magazine*

"The Kano model is a powerful concept in the world of quality and customer satisfaction. Lance Coleman's book provides a very helpful introduction, illustrated with many real-world examples that help the reader get beneath the surface of the model. All types of companies, including healthcare organizations, can use this book to verify that they truly understand what their customers (or patients) need and the things that will surprise and delight them!"

—Mark Graban, author of *Lean Hospitals* and *Healthcare Kaizen*, Consultant, Speaker

"With engaging stories and solid examples of how it can be applied, this book shows how to put the Kano model to practical use along with Hoshin Kanri and Quality Function Deployment. A good roundup of Lean, process improvement, control, and planning makes this book a great guide to putting the customer at the heart of the organization."

—Melvyn Thornley, M.Eng, MBA, CQP, CSSMBB; 2014 New Zealand Organization for Quality Vice President; Managing Director, Thornley Group

"This book brings about the understanding not only of customer satisfaction as most of us understand it but something much deeper than that. Let's call it 'Meta Satisfaction'—the satisfaction of experiencing satisfaction. The message in the book is epitomized by a reflection on the conversation Lance and his wife [had] with a cabin attendant on a cruise ship. When offered praise and a tip for a service rendered the cabin attendant replied, 'No tip, thank you, it is my pleasure to serve you.' Now we are talking Quality."

—Paul Harding, MSIA; Executive Director, South African Quality Institute

T0341107

The
Customer-Driven
Organization

The
Customer-Driven Organization

EMPLOYING THE KANO MODEL

Lance B. Coleman, Sr.

CRC Press
Taylor & Francis Group
Boca Raton   London   New York

CRC Press is an imprint of the
Taylor & Francis Group, an **informa** business

A PRODUCTIVITY PRESS BOOK

CRC Press
Taylor & Francis Group
6000 Broken Sound Parkway NW, Suite 300
Boca Raton, FL 33487-2742

© 2015 by Taylor & Francis Group, LLC
CRC Press is an imprint of Taylor & Francis Group, an Informa business

No claim to original U.S. Government works

Printed on acid-free paper
Version Date: 20141020

International Standard Book Number-13: 978-1-4822-1710-0 (Paperback)

Visit the Taylor & Francis Web site at
http://www.taylorandfrancis.com

and the CRC Press Web site at
http://www.crcpress.com

Dedication

I dedicate this book to my wife and family. My wife of 27 years, Lorraine, continues to be a source of love, support, and inspiration in all that I do. My four children, Larissa, Lauren, Lance Jr., and Latrice were a joy and blessing to raise; now as adults I am proud to call them friends. Finally a shout out to my dogs Auggie, Leo, and Shii whose boundless exuberance and joy upon my simply entering the house, bring a burst of joy into even the most dismal of days that we all sometimes face.

Dedication

Contents

Preface

Who wouldn't want to work more effectively, more efficiently, more effortlessly? Who wouldn't want to create value, reduce waste, and increase customer loyalty? How can we do this?

How often have you heard someone say, "I don't remember the last time I had good customer service"? How often have you heard a version of the expression, "One satisfied customer may tell one person but a dissatisfied customer will tell everyone they know"? The flip side of that expression is, one extremely satisfied or "delighted" customer will also tell everyone they know. So when you think about it, "delighting" customers just makes good business sense. This concept was the crux of my presentations during the 2013 American Society for Quality (ASQ) Lean and Six Sigma Conference as well as the 2014 ASQ World Conference on Quality and Improvement, and is now the subject of this book.

I was first introduced to the Kano model while attending a session at a prior Lean and Six Sigma conference by Lean guru and author of *Gemba Walks for Service Excellence*, Bob Petruska. During the presentation, Bob explored how the practice of providing exceptional customer service can drive business success and gave examples across a wide range of hospitality and service industries including hotels, auto repair shops, airports, and a campsite provider. Bob's presentation took us on a virtual journey from Texas to Colorado to Ohio to Japan and beyond by providing concrete and inspiring examples of what he was discussing.

In framing the concept of customer service (exceptional, good, and poor), Bob introduced the concept of Kano analysis developed by Professor Noriaki Kano. I was both captivated by his enthusiasm for the subject, often punctuated by an "Awesome!" as he made a particular point, as well as intrigued by the simplicity and effectiveness of this tool. As I sat in the session, I started to speculate about the untapped potential that may still be there. I began to see how the concepts of internal versus external customers could also be explored and began to think about how the Kano model might be successfully used in my manufacturing environment.

Every two to three years at any major Lean or Lean Sigma conference, there will be at least one session on the Kano model. Since first sitting in Bob's session, I have reviewed a number of conference session proposals, and sat in on several classes relating to the Kano model. They all seem to have one thing in common: examples of Kano model use always seem to be taken from the service industry. I continued to ask, "Why can't this tool be used with some modification in manufacturing as well?" After successfully using aspects of the Kano model at my own job, I decided to propose a conference session based on using the Kano model in a manufacturing setting (how, why, and when to use) to share my successes. I also wanted to explore the concepts of internal and external customers with session attendees. The selection process for presenters at the ASQ Lean and Six Sigma conference is very competitive and I was thrilled to learn that I had been selected to present. The session was well received by the attendees and afterward Michael Sinocchi from Productivity Press came up to me and asked me if I would be interested in writing a book based on this topic. "Hmmm," I thought, "maybe I am on to something after all." Michael certainly thought so. I hope you do as well after you finish reading this book.

Acknowledgments

Did I mention that my wife Lorraine is also a consultant, lecturer, and published author? If I do manage to maintain a conversational tone throughout this book, if the stories that I share are both engaging and thought provoking, clearly illustrating the points I am trying to make, it will be thanks to what I learned from Lorraine. If I did not manage to do any of those things, then the fault lies solely with myself.

Chris Hayes, Dr. Terra Vanzant-Stern, Scott Smith, Bob Petruska, and Duke Okes have become friends as well as mentors for me in traveling along my Lean and quality journey. My thanks to them, as well as Imelda Hernandez and Mark Nestle for their feedback on various sections of this book. Again, any falling short in the presentation of the principles discussed in this book lies solely at my own feet.

My further appreciation to Jonathan Clark, whose writing workshop that I took just as I was beginning to work on this book, gave me many of the tools that I used herein to deliver a much stronger manuscript at a much faster pace than I would have ever expected.

Thanks to Michael Sinocchi and the team at Productivity Press for selecting me to write this book and for guiding me through the process. And finally, but most important, my thanks to Dr. Noriaki Kano for developing this powerful tool in the first place.

About the Author

Lance Coleman has over 20 years of leadership experience in the medical device, aerospace and defense industries and presently serves as a quality engineer and Lean program leader at The Tech Group in Tempe, Arizona, USA. In that capacity he also manages the site corrective/preventive action, customer complaint and internal quality audit programs. Lance earned a degree in electrical engineering technology from the Southern Polytechnical University in Marietta, Georgia, USA, and is an ASQ senior member, as well as Certified Quality Engineer, Six Sigma Green Belt, Quality Auditor and Biomedical Auditor. He is an instructor for the ASQ CQA exam refresher course, presently serves as the ASQ Lean Enterprise Division Education Committee chair and is the editor of the ASQ Audit Division newsletter. As the founder of Full Moon Consulting since 2012, Lance has delivered live, online and blended learning training on the topics of Lean implementation, risk management, and quality auditing across four continents.

For questions or comments, you can contact Lance Coleman at lance@fullmoonconsulting.net.

Introduction

Like so much in Lean philosophy, the Kano model is really a way of thinking that can be applied to everyday situations. Then these thoughts are given concrete application through the use of varied and diverse tools that we discuss throughout this book. To be clear, in writing this book I am not trying to recruit you to any specific way of thinking, but rather want you to open your minds to alternative ways of thinking about how we relate to the customer, our coworkers, our bosses, and varied individuals with whom we interact every day. The customer-driven organization is merely one road on the path to delivering delightful service and reaping the resulting rewards.

In *Gemba Walks for Service Excellence*, Bob Petruska displays the power of delivering delightful service through many service business examples from the United States and around the world. This book is written to show how to use quality, Lean, and Six Sigma tools to implement the Kano model successfully in diverse organizations, and to show how to develop your organization into one that is forward facing and customer focused. A second goal of this book is to show readers how to take these same powerful concepts and lessons learned from the service sector and apply them on the manufacturing floor. In the following pages, the book includes chapters devoted to the following topics: Introducing the Kano Model, Shifting Focus, Planning Tools, Deployment Tools, Metrics and Monitoring, Root Cause Analysis, Project Management, and finally, Putting All the Pieces Together, which contains step-by-step examples of how to deploy the Kano model in your organization. The book uses case studies and real-life examples throughout to illustrate key points.

I attempt to show how deployment of the Kano model in a manufacturing setting can lead to internal efficiencies and heightened morale and drive business improvement as it does in the service sector. The concept of the voice of the customer (VOC) is introduced along with those of value, value proposition, and value stream. These concepts are all explored within the framework of the Kano model as used in diverse settings. An additional emphasis is placed on identifying internal as well as external customers while recognizing both their implicit and explicit needs. Last, I discuss how the Kano model might be implemented successfully and briefly touch on the psychology of receiving and providing exceptional customer service.

Throughout this book I use the general term *organization.* However, I could just as easily use the terms *manufacturing company, service provider, nonprofit organization,* and, in many cases, even *individuals.* The Kano model is a broad concept with an expansive potential for successful usage. This book is meant to serve as a practical field guide in the deployment of Kano analysis and the Kano model in your work and home settings. It also teaches readers how to use Lean, quality, and Six Sigma tools to drive toward the goal of delivering exceptional value and hence delightful service to both internal and external customers.

Right Brain versus Left Brain

It has been said that the right side of the human brain is where logic and method are centered and the left side of the brain is where creativity holds sway. For certain individuals, one side or the other may be predominant with regard to how they interact with life, or process experiences, but everyone has aspects of both the engineer and the artist in their psyche. So what does this have to do with delightful service?

Have you ever done something for someone and been pleasantly surprised at the intensity of the response as the good deed may have been much more beneficial to the recipient than you had ever imagined? Ever worked on a project and delivered more than expected, ahead of schedule? Just "knocked it out of the park"? These experiences made you and everyone involved feel good, didn't they? In this book we explore in depth how delivering delightful service can drive continual improvement, customer loyalty, and subsequently, business success. During these discussions, however, we must not lose sight of the intangibles, namely, that delivering delightful service feeds both the left and right sides of the brain. It not only makes business sense, it feels good, too.

Finally, I would like this book to serve as a reminder to continue to ask yourself the questions: am I providing the best possible value to the people with whom I interact on a daily basis, and if the answer is no, then why not?

Chapter 1

Introducing the Kano Model

The Kano model is a theory of product development and customer satisfaction developed in the 1980s by Professor Noriaki Kano, a student of Kaoru Ishikawa (of Ishikawa diagram fame). The development of the Kano model came out of a scientific study investigating the varying definitions of quality and their significance.[*] From this study came the recognition of two distinct aspects of quality—objective (physical fulfillment or compliance with specifications) and subjective (end-user satisfaction)—as well as, their correlation. Next five broad classifications of quality elements were defined that reflected the customer experience.

> *Attractive Quality Elements:* Elements that when fulfilled provide satisfaction but are OK for the customer when not fulfilled.
> *One-Dimensional Quality Elements:* Elements that result in satisfaction when fulfilled and dissatisfaction when not fulfilled.
> *Must-Be Quality Elements:* Elements that are absolutely expected but result in dissatisfaction when not fulfilled.

During the study, the first three classifications were seen to be the most common scenarios but the two below were also found to be possible.

> *Indifferent Quality Elements:* Elements that neither result in satisfaction or dissatisfaction, regardless of whether they are fulfilled.
> *Reverse Quality Elements:* Elements that result in dissatisfaction when fulfilled and satisfaction when not fulfilled.

---

[*] Kano et al., 1996.

This information was all plotted on a two-dimensional Kano diagram which, all of these years later, has changed very little. The power of this study was that it allowed the research team to predict certain behaviors. By creating surveys based on the established quality element classifications the research team was able to draw conclusions about the buying habits and expectations, both implicit and explicit, of different segments of the population. It was then realized that this information could be critical in allowing organizations to make strategic design development and marketing decisions as part of their innovative process.

Today, as in the 1980s the Kano model is used as a means to:

- Prioritize critical to quality characteristics (those most important to the successful function or fulfillment of purpose) of a product or service as defined by the customer
- Identify implicit as well as explicit customer needs.

Kano analysis looks at customer service and the benefits of delivering exceptional value to the customer through the vehicle of delightful service. The Kano model still recognizes five states: indifference (characteristic does not matter to customer), nonperforming (failure), basic "must haves" (cost of entry into the marketplace), performing (more is better), and exciting or delightful service (surprises and delights the customer). Three of these states are shown visually in the Kano diagram in Figure 1.1. The x-axis of the Kano model is labeled desired characteristics and the y-axis is labeled customer satisfaction. Use of this model stresses the fact that we need to provide desired characteristics as a priority and that true success is measured against customer satisfaction.

Nonperforming speaks for itself; we obviously want to avoid this category whenever possible. Per the Kano model, a company providing "basic must haves" is meeting basic customer requirements. It is meeting minimum customer expectations. Without providing "basic must haves" to the customer, companies wouldn't stay in business and employees wouldn't keep a job. Basic customer requirements are uncovered through market research, surveys, interviews, focus groups, customer feedback, and other similar activities.

Performing occurs once an organization hits its stride and begins to move along the performing continuum; improving over time through economies of scale, continuous improvement efforts, and other factors. Evidence of performance could include any or all of the following: the service or product costs

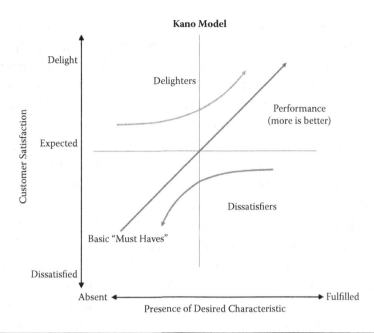

Figure 1.1

less, is delivered more swiftly, and has incorporated improvements. It is important to note, however, that even when minimum requirements remain the same, minimum expectations begin to rise as performance improves. Information on performance expectations is derived by field visits, customer feedback, external failure reports, internal failure reports, benchmarking, and other similar activities.

One concept that may seem initially hard to grasp, but that makes sense once you think about it, is that customer satisfaction does not necessarily mean customer loyalty. Customers can be satisfied with a product or service and still go elsewhere for the sake of convenience, price, or some other reason. The goal of every organization should be to transition through the stages of innovation development, branding (name recognition), customer satisfaction, and finally on to customer loyalty. Customer loyalty is what drives active referrals and increases business. That is what organizations should be seeking. Using the Kano model sets this target as an up-front goal.

Key Point

Delivering better and better performance "raises the bar" on expectation, yet does not guarantee customer loyalty.

Delightful service surprises and excites customers by exceeding their expectations. Sometimes delightful service means bringing a project in early or under budget, or with higher than expected quality. Sometimes delightful service means anticipating customer needs that the customer might not even realize that he or she has at that point in time (innovation). How do we anticipate customer needs? Through market research combined with trend analysis, customer feedback, benchmarking, and by concentrating on the end purpose of the service or product. We can best examine the end purpose of a good or service by looking at the customer's implicit, as well as explicit needs, which is a topic that we cover in the next chapter.

One thing that sets delightful service apart from the other states recognized in the Kano model is its positive effect on the giver as well as the recipient. Who hasn't felt a rush when a customer is pleasantly surprised and excited when they receive better than expected service? Delightful service makes everyone feel good. Delightful service makes you want more. Delightful service drives continuous improvement and business success.

| Key Point | *Delightful service means exceeding customer expectations by anticipating their current and future needs* |

If we are going to talk about delightful service, we must also look at defining the concept of value. Value must come from the customer's perspective. If the customer is willing to pay for something, then it has value. You could also add that a transformation must take place and that the process was done right the first time. (A repeated step is 100% waste.) Think of the customer–supplier relationship as a person looking at a black box. The customer inputs an order along with payment and expects to receive a good or service that meets her requirements and is delivered on time. The customer really doesn't care what goes on inside the black box (what the supplier needs to do) to make that happen.

An example of a value-adding step in manufacturing is inserting a bolt or screw that is necessary to hold a piece of furniture together so that it can perform its function. Final inspection of the sofa prior to shipping would be considered a non-value-adding step, because it does not add anything to the finished product. The inspection could still be considered important as it helps keep defective merchandise from getting to the customer, yet it is non-value-adding because as an individual line item, the customer would not be willing to pay for it. After all, they are purchasing a sofa and not an inspection service. The inspection would be considered a cost of doing business that the supplier factors into the pricing.

A value proposition is how an organization positions itself to stand out from its competition. It is those attributes of the product or service that a company provides which make that company unique in the marketplace, the reason(s) that the customer chooses company "A" over company "B." Internally this can mean comparing your services provided against your predecessors or peers. Are we proactive in our approach, constantly looking to enhance our skills and improve our methods? Are we tracking appropriate metrics in order to assess our level of performance against expectations and goals, as well as against others performing similar activities?

The value stream is the aggregate of all activities, information, and materials needed, both value-adding and non-value-adding, to deliver a good or service to a customer.

Key Point

An important or even critical process step can still be considered non-value-adding from the customer's perspective

Case-In-Point: Power of Kano in Action

There is a manufacturer of industrial kitchen appliances who embraced the philosophy of Kano with outstanding results relating to its line of industrial ovens. In this market, features were fairly consistent across brands. Companies jockeyed for market share through a combination of product cost, quality, and reliability, as well as company reputation. Eventually, someone at this company got the idea to go to key customers who were using their ovens, watch the ovens in use, and inquire as to things that the users didn't like or had to work around. They also contacted customers of their competitors' ovens to make similar inquiries. Eventually this company had a list of annoyances, dislikes, hazards, and needed workarounds that were consistent across companies, regions, and brands, when it came to industrial ovens.

The company then took this list to their design team and told them that it was a priority to design these issues out of the next iteration of their ovens. The resulting product was successfully beta-tested at a client of theirs and then released nationwide to great success. In fact, the product launch was so successful that the company greatly increased its market share within a year's time and continued to increase market share incrementally for years afterward. It is significant to note that typically in this industry, industrial ovens are used until they can no longer function; then they are replaced or upgraded at this time. The new oven was so sought after that a significant number of companies replaced their existing ovens even prior to the conclusion of their own ovens' life cycles.

This story is an example of a company aggressively pursuing an opportunity to delight their customer and reaping substantial benefits from doing so. Because the issues encountered were consistent across brands and price ranges, the end users had come to expect certain annoyances, needed workarounds, and even hazards when using industrial ovens. They weren't expecting any changes any time soon and so were delighted with the redesign of the oven and its greatly improved performance. So by observing and strategically asking questions, the company was able to identify and meet customer implicit needs, thereby providing customer delight and increasing market share as a result.

Case-In-Point: Delightful Service

C.I.P. 1

My wife Lorraine and I went on an Alaskan cruise for our fifteenth wedding anniversary. Whenever we left the cabin, our beds were made and turned down with candies put on our pillows. This was not the delightful part. This was somewhat expected. Once we forgot something in our cabin and headed back before we got to the stairway at the end of the hall. We were shocked to find that the deck attendees had somehow already been in our cabin and "done their thing" without our even seeing them leaving as we were coming back down the hallway.

As the cruise progressed, my wife and I made a game of coming back unexpectedly on occasion to see if we could catch sight of one of the "elves," as we began to refer to them, tidying up our room but never did. Toward the end of the cruise I did finally see one of the attendants (not coming out of our cabin though). I greeted him and tried to give him a tip which he politely refused. When I thanked him for the excellent service that he had provided, do you know what he responded? "It is my pleasure to serve you." Wow. How cool is that? Pleasure in service. What an amazing concept that helps deliver memorable, or dare I say it, delightful service. Twelve years later, I remember with great fondness that trip and that conversation.

C.I.P. 2

Years ago, at the start of my career, I was brought in to revitalize an inadequately performing internal audit program that was causing customer complaints. By the time I arrived, the situation had gotten to the point where the on-site representatives for two different customers were threatening to shut down production lines if things didn't improve. The procedures and work instructions were unclear and often contradictory. Audits weren't being done per the schedule and those that were being conducted were done in a slapdash fashion. On more than one occasion a customer representative had gone into an area that had just successfully passed an internal audit and found multiple (and sometimes major) noncompliances. You don't want to see this type of lack of control in any facility, especially not in one that manufactured rocket propellant! Rewriting procedures and scheduling and conducting audits was easy. Rebuilding the trust of the on-site representatives was the hard and most important part. What could I do?

Once I had begun to rebuild the infrastructure of the internal audit program, I asked my manager if I could copy the customer representatives on the audit results in addition to the audit notifications. I was advocating total transparency. He was hesitant at first, so I asked him, "At this point, what do we have to lose?" He agreed and I let our customers know that I was about to start implementing the new schedule and asked if they would be interested in receiving our audit results. They were both very pleasantly surprised by the offer and did ask to be included on the audit report distribution.

The issues with the audit program were resolved but, more important, the relationship with our customers was greatly improved. The frequency of customer audits and the subsequent time needed to support them was reduced. Sample size on outgoing lots was reduced which subsequently reduced cycle times. We even reached a point where the customer representatives would call me and let me process an internal finding when they found a minor violation during one of their audits, and copy them on the response, thereby cutting down on the amount of paperwork for everyone.

So in this case, by delivering more than what was expected in the form of transparency of the audit program, customer relations were improved and the bottom line positively affected by shortening cycle times and reducing the amount of resources needed to support customer audits.

Case-In-Point: Performing

C.I.P. 1

Does anyone still use or even remember WordPerfect word processing or Lotus 1-2-3 spreadsheet software? They each dominated their share of the home software market before Microsoft got through with them. Microsoft is an excellent example of iterative improvement over time, innovative business strategies, and leveraged technology used to dominate the marketplace. When Microsoft saw how popular Apple computers were with graphic icons and pull-down menus they developed their own Windows operating system with similar features. They then struck a deal with non-Apple computer makers to incorporate this very popular operating system in the new computers that they sold. The next step was to develop their own word processing and spreadsheet software packages and bundle them for a relatively small fee with the new Windows operating system. This way millions of people each year had access to, and became comfortable using the Microsoft products instead of their competitors'. These techniques brought Microsoft from two guys in a garage a little over three decades ago to a Fortune 37 company in 2012. Eventually, Microsoft Word and Microsoft Excel became dominant in their niches.

Continually looking to expand beyond the realm of computers where they got their start, Microsoft products are found in many places where there is software such as web search engines (bing), game systems (Xbox), and smartphones (Windows Mobile). However, now, much more than earlier in their lifecycle, they are more likely to arrive late to a proven market and then leverage their existing technology to carve out and expand their own niche. Microsoft invests top dollar to get top new talent and is always striving to develop new technologies, but is just as likely to purchase a proven software that complements what they already do well (e.g., Visio).

So by continuing to improve, benchmarking, leveraging existing technologies, and innovative business strategies, Microsoft has come to dominate many aspects of the software market. A perfect example of how continually improving performance leads to success.

But sometimes performance excellence isn't enough.

C.I.P. 2a

Kodak created the camera and film market and then dominated it for years. They achieved economies of scale and state-of-the-art film and camera-making technologies. Then they made one "little" mistake. They passed early on digital camera technology. More than 100 years old and a fixture on the Fortune 500 list for years, Kodak filed for Chapter 11 bankruptcy protection in 2011. Now late to the market, Kodak is focusing on digital photography and digital printing to try to drive new business as they emerge from bankruptcy. So what was the difference between Microsoft and Kodak's results? Both companies achieved economies of scale and pursued the latest technologies along with continuous improvement efforts. The primary difference is that Microsoft successfully included external feedback into their business model.

C.I.P. 2b

Many years ago my wife Lorraine and I owned a small kiosk in Crabtree Valley Mall in Raleigh, North Carolina, where we sold African American books, greeting cards, art, and gift items. Despite the small 24-square–foot space, we were very successful comfortably supporting a family of six. As the largest mall in the area, many tourists came through, bolstering our business during the off-peak season and many of our recurring local customers became friends. We occasionally traveled to craft shows and cultural events to keep our inventory fresh and customers happy. To further the unique aspects of our business, we developed our own product line. Our business had even been featured in several of the local weekly periodicals. Life was good.

Within an 18-month period, the mall Barnes & Noble substantially increased their offerings of African American literature (wonder where they got that idea), Sears launched their African American greeting card line, and a new African American art gallery opened in a nearby suburb. Despite the business improvements that we had made and the standard of excellence that we had maintained, we lost substantial business. Our inventory portfolio was outstanding and our customer service was excellent. Now, if we could only have competed pricewise with Barnes & Noble and Sears! We actually retained most of our customers, but they just didn't spend as much or as frequently with us. Due to a lack of resources, we were slow to create or seek other new and innovative product lines. So what happened next?

Did I mention that I work as an engineer now?

Chapter 2

Shifting Focus

… conformance to requirements will not ensure quality. Zero defects is not Quality. Quality is much more complex than zero defects …

W. Edwards Deming
September 22, 1987

In this quote, Deming was talking in part about the need for leadership to forecast the market, to look at new ways of doing things, and to identify and meet the unspoken needs of the customer. Innovation adds value to the supplier–customer relationship whether through new technology, new materials, or new methods. Innovation starts with market awareness, with knowing what's possible. It is in this way that organizations can begin moving to meet the implicit needs of their customers.

We live and die by information in this society. Despite this fact, we often don't communicate our needs adequately, as organizations, as individuals, in business, and in our private lives. There are three elements that are critical to providing exemplary service: communication, observation, and analysis. When initiating communication, these three rules should be followed:

- Make sure that the message is clear and concise.
- Make sure that the message is received.
- Use the appropriate communications medium.

When receiving communications the following rules should apply:

■ Confirm that you have received the entire communication and all attachments.
■ Read the entire communication carefully (twice) before responding.
■ Clarify any questions that you have before responding.

It is also important to be able to "read between the lines," provide context, identify nonverbal cues, and recognize trends in diverse data when initiating or receiving communications. It is even more important to analyze these data properly so that you can respond most appropriately. This ability separates organizations that delight their customers from those that merely satisfy them.

Have you ever had a day where you were frustrated because you worked hard but didn't get a lot done? Conversely, have you ever had a day where you were humming on all cylinders, the customer was delighted with you, and you felt great? People want to do a good job; they want to succeed at their tasks. The Kano model provides a framework that organizations can use to put people in a position where they can succeed in providing exceptional service, delight the customer, and drive business success. So how is this accomplished? It is accomplished by shifting our focus.

Meeting requirements should be the starting point for an organization to consider itself successful. Value should be treated as its own intangible line item and be aggressively sought out for every project, every good, and every service. When management reviews a good or service ready for delivery, there are several questions that need to be answered:

1. Did we meet requirements (specifications, cost, delivery date)?
2. What is our value add (good or service beyond the technical specifications that our customer will find of value)?
3. Are we clearly, and with discrimination, discerning the voice of the customer?
4. Are we looking to add value at all phases of our operation?
5. Are there adequate feedback loops so that we can quickly make adjustments to how we do things when necessary?
6. Is our response time adequate to the needs of our business?

This attitude must flow down from the top executive to the front-line employees. Transformation of this kind is not accomplished through exhortations or through writing standard operating procedures. It is achieved through communication of values as well as organizational goals and objectives. This practice, combined with formal training in the tools that the employees will use to accomplish their objectives, will in turn support the organizational strategic vision.

The *Voice of the Customer* (VOC) can be defined as the collective insight into the wants, needs, expectations, and perceptions of the customer derived through direct and indirect questioning. This insight can also be derived from observation and data analysis. It is used to match as closely as possible the product or service developed by an organization to its customers' needs, both explicit and implicit. An important question to ask while going through this determination is how does the customer perceive value? What are his priorities? What is the hierarchy of his needs? Hoshin planning and quality function deployment (QFD), which we cover in the next chapter, are two of the tools that can help us to make that determination. For now, let us say that the customer perceives value in how well his needs and expectations are met. Thus, it is very important for us not only to receive the voice of the customer, but to be able to translate it accurately as well.

There are two types of customer needs: explicit and implicit. Explicit needs are fairly straightforward. They are communicated directly and expressed either verbally or in writing. Explicit needs may take any of the following forms

- Contracts
- Drawings and specifications
- Customer feedback
- Correspondence
- Statements of work (SOW)
- Verbal communication

Key Point

It is not enough to perceive the voice of the customer. It is critical to translate that voice into needed actions.

Implicit needs are a little trickier to determine. Have you ever been house or car shopping and your realtor or car salesman shows you a model with everything that you asked for, within your budget, and yet still you hesitate? That is because there is an unspoken or implicit need, that you may not have been aware of, that hasn't yet been met, despite having all of the

Kano Model Sample Needs Survey*

How would the customer feel if the need WAS addressed?
(Positive)
How would the customer feel if the need WAS NOT addressed?
(Negative)
The customer has 4 choices in response to each question:
- I'd like it
- It is normally that way (the feature negative/positive is expected)
- I don't care
- I wouldn't like it

> * Thanks to Bob Petruska
> of Sustain Lean LLC

Figure 2.1

other positive characteristics that you desired. Not only are implicit needs unspoken and sometimes difficult to determine, they are often more important than explicit needs. Sometimes an implicit need may be tied to the purpose or function of the product or service. Sometimes it may be tied to an emotional need that the right side of the customer's brain has discarded as not very important. At other times, implicit needs may be tied to cultural differences, increasingly important as the world becomes more global. Some methods for determining implicit needs are seen below. A sample implicit needs survey is shown in Figure 2.1

- Customer feedback
- Correspondence
- Verbal communication
- Customer/market surveys
- Focus groups
- Internet portals

Key Point

1) *Sometimes implicit needs are more important than explicit ones.*
2) *Sometimes the customer may not even realize that he has implicit needs or what those needs specifically entail.*

Customer feedback is an important part of process monitoring and improvement. Feedback can be gained either passively or actively. Passive (on the part of the organization) feedback is presented by the customer, usually in the form of correspondence, complaints, change requests, supplier score cards, audit reports, or some combination of these documents. It typically addresses explicit needs, although sometimes it can point to implicit needs. Active feedback is solicited by the organization in the form of surveys, phone calls, observations, correspondence, visiting Internet portals, and other similar activities. Active feedback is better at capturing implicit needs but may identify explicit needs as well.

Feedback can be positive, negative, informative, or interrogative. Positive feedback may take the form of supplier score cards, supplier awards, correspondence, renewed/expanded contracts, audits, and other reports. Negative feedback could take the form of complaints, supplier score cards, customer actions (second supplier brought in), audit reports, and other correspondence. Informative feedback might take the form of changes seen in a product or service that may not necessarily be out of specification. It could be a notification of upcoming changes in need from the customer. Interrogative feedback would usually be questions about capacity (can you produce more, quicker, modifications, etc.).

Key Point

Customer feedback can be gained actively or passively. It can be
1) *Positive*
2) *Negative*
3) *Informative*
4) *Interrogative*

Wikipedia defines *customer* as "the recipient of a good, service, product, or idea, obtained from a seller, vendor, or supplier for a monetary or other valuable consideration." Two types of customers must be taken into account—external and internal—and they each have their own explicit and implicit needs. An *external customer* can be thought of as the one that receives the finished product or service and that pays the creating organization. An *internal customer* can be thought of as the next step in a process or series of processes, or a department or function that requires input from another department, function, or facility within an organization.

So who are external customers? Some examples of external customers are

- Original equipment manufacturer (OEM)/design authority
- End user
- Retailer/end distributor
- Other sites

Some explicit needs that an external customer might have are

■ Technical requirements
■ Cost
■ On-time delivery
■ Established quality management system (QMS)
■ Risk management

Implicit needs that an external customer might have are

■ Functionality
■ Improvement
■ Error detection
■ Aesthetics

Now how about internal customers? Here are some examples of internal customers

■ The next step in a process (or the individuals performing the next step in a process or series of processes)
■ Other departments
■ Site management
■ Direct reports

Some explicit needs for internal customers are

■ On-time delivery
■ Error-free delivery
■ Proper documentation

Some implicit needs for internal customers are

■ Error detection
■ Risk identification
■ Improvement

If delivery of a good or service on time, to specification at the promised cost (internally cost could be replaced with use of resources) is the "price of entry" to the marketplace, how do we add value for our customers? Lean, Six Sigma, and other continuous improvement efforts can lead to more efficient operations which in turn can reduce cycle times, lower costs, and increase quality on the good or service delivered. Supplier monitoring and development can lead to lower costs in raw materials, supplies, components, and subassemblies. These cost savings can be shared with the customer. We can also minimize the use of resources to get the same result.

As the subject matter expert (SME) in what they do, subcontractors can often make design change recommendations, catch errors, identify unaddressed risks, and recommend alternative materials or methods. This is accomplished through establishing communication channels that encourage information exchange and collaboration.

Case-In-Point: Communication and Context

So what do we mean by "appropriate method" or mode of communication? This would be the method of communication that the recipient is most comfortable with, as evidenced by how they initiate contact. For example, if a person responds to your e-mail with a phone call, the phone is probably her preferred method of communicating. The converse would also be true. However, even when trying to make recipients more comfortable in their communications, certain norms of professionalism should hold true. If communicating by phone, be sure to follow up with a communication in writing confirming what was said. Or conversely, send an e-mail then follow up and discuss it over the phone.

Two additional caveats: text messaging should not be used in professional communications that may become part of a project/company record, and in cases of a legal nature, "snail-mail" is often still an important part of the communications chain.

Context is important because it can provide the information that an organization needs to respond appropriately to a given situation. For example, an organization gets a rating of "Good" from one of its customers. The response might differ depending on whether the last five ratings for that organization were, "Fair," "Good," or "Excellent." It might also differ if other suppliers of similar goods or services were typically rated as "Fair," "Good," or "Excellent."

A customer brings on board an alternative supplier to supply the same good or service that your organization provides, but your order volume does not change. One reason may be concerns about your organization's quality, delivery time, or costs. Another could be an anticipated upswing in need of the customer for the part that your organization may not currently have the capacity to fill, or that the customer may believe that your organization does not have the needed capacity: two totally different scenarios, requiring two totally different responses. Without context, a disastrous decision could be made.

Case-In-Point: Explicit versus Implicit Needs

Let us take a look at the case of a carpenter who has been asked to make a cane for a customer who is 6'2" and weighs 220 pounds and is known to be a snappy dresser. The specified cost, materials, finish, and dimensions would be the purchase requirements or explicit needs. In addition to the requirements included in the purchase order, the cane would need to be sturdy enough to support a 220-pound man and long enough for a 6'2" tall man to use. Even if the requirements explicitly stated in the purchase order are met, if these two implicit needs aren't met, then the cane would not be usable. Similarly if all of the other needs are met and the cane is not fashionable, then the customer may not buy it, or if he still buys it, he would not be happy. It is important to assess whether the implicit needs of function are met in addition to the specified requirements.

Explicit	Implicit
Cost	Sturdy enough
Dimensions	Fashionable
Material	Nonslip
Finish	
Footer	

Another example of the need to match implicit with explicit needs is a screw. The thread can meet an ANSI (American National Standards Institute) specification that calls out thread width and pitch, and meet requirements for length and diameter (all explicit needs), but if the screw won't screw into the mating part (implicit need, functionality) then it is worthless. This is a case where the implicit need is at least as important as the explicit requirements.

Case-In-Point: Voice of the Customer

VOC can be obtained directly from the customer, from the market as a whole, or from the end users of the good or service (if not the customer). Sometimes a need can be implicit simply because the customer did not think in terms of your organization providing a particular good or service. VOC should be acquired throughout all phases of a project. Careful observation and effective questioning is also very important to determining VOC as sometimes the customer may not be aware of all of their needs. For both internal and external customers the process of gleaning information from their voice is very similar.

External Customer

- The initial explicit requirements are stated in the contract, purchase order, quality agreement, or some combination of the three.
- The customer will provide ongoing feedback through correspondence, verbal communications, audit reports, supplier score cards, and customer complaints.
- Going to the end user's location (or watching a video if travel is impractical) and seeing how the good or service is used is another good way to ascertain customer needs both explicit and implicit.
- Organizations can be proactive by sending out surveys and developing keen observation skills. Strategically worded questionnaires may provide insight into future opportunities.
- Attending trade shows and reading trade periodicals allow an organization to keep abreast of the latest technology and current trends in the market. This awareness can lead to ideas for new business opportunities with customers, or ways to add value to existing customer relationships.

Internal Customer

- The initial explicit requirements may be found in work instructions, technical specifications, and job descriptions.
- Going to the next step in the process and seeing how the output from your process is used is often helpful in determining how processes can work more effectively and efficiently together.

- Internally, individuals can be proactive by communicating with stake-holders and process owners for the process that their operation feeds into and asking what they are doing well, and what can be done better, also, by implementing Kaizen, Lean, and Six Sigma initiatives and continuing to look for improvement opportunities within their own process. Tools such as process mapping aid in uncovering waste and areas where improvements can be made.
- Attending professional conferences and reading industry periodicals may be a way to learn methods and technologies that will make internal processes more effective and efficient.

Case-In-Point: Internal Customers

If the process [activity(ies)] that you are engaged in provides a good (raw materials, components, subassembly, manuscript) or service (inspection, test, document review) to the next process in a series of processes, then that next process is your customer. Your customer wants the good or service delivered per requirements, on time, while utilizing the allocated level of resources to do so. Some ways to add value might be to reduce start delays, processing time, and errors, as well as the resources needed (one example would be going from two people down to one to perform a process).

An internal customer can also be the person(s) above or below you on the organizational chart. Organizationally, managers have obligations to three customers, (1) their superiors, (2) the organization as a whole, and (3) their direct reports. The site/organization sets goals ("purchase require-ments") then provides salaries and resources ("purchase price") and expects goals and objectives to be met.

One management pitfall to avoid is that sometimes managers are better at addressing the needs of the organization than they are the needs of their direct reports. Managers have an obligation to provide adequate instruction, training, resources, support, and professional development ("purchase costs") of their direct reports. Otherwise they will take their "business" elsewhere by transfer-ring to another department or leaving the company. Studies have shown that happy and engaged employees are more productive, take less time off from work, and remain with employers longer. Happy and engaged employees are those who have their implicit as well as their explicit needs met.

Chapter 3

Planning Tools

Often when the Kano model is discussed, the conversation is framed in terms of what happened and why certain actions led to success, but not how the organization was able to achieve its desired future state. For the sake of this book I have decided to separate the tools into three categories: planning, implementation, and monitoring/feedback. In each category, we go over the simplest tools first then progress to the more complex ones. In this chapter we talk about the various quality, Lean, and Six Sigma tools that we can use to apply the concepts of the Kano model in practice, during the planning phase of a project, business, or nonprofit launch. The tools we cover are seen below.

- 20 Keys Assessment
- Inverse Ishikawa Diagram
- Hoshin Kanri X-Matrix
- Quality Function Deployment

20 Keys

Prior to embarking on a plan to improve customer service (or any improvement plan for that matter), it is important to assess the current state of your system and to determine where it is you would like your organizational system to be in the future. Professor Iwao Kobayashi[*] did extensive research

---

[*] Kobayashi, 1995.

on, and personally experienced, the rise to world class of numerous Japanese companies during the 1980s and 1990s. The 20 keys are an assessment and improvement monitoring methodology created by Kobayashi in the 1990s, based on this experience. During this research he discovered 20 critical focus areas for a company's sustainable and holistic development. These 20 focus areas became known as the 20 keys. This approach embraces the philosophies of Lean, Six Sigma, technology advancement, continuous improvement, and worker empowerment. The goal in developing this tool was to highlight the 20 "keys" that were crucial in developing a world-class company and provide a pragmatic approach to the pursuit of excellence.

The original 20 keys were

Key 1:   Cleaning and Organizing
Key 2:   Rationalizing the System/MBOs
Key 3:   Improvement Team Activities
Key 4:   Reducing Inventory (Shortening Lead Times)
Key 5:   Quick Changeover Technology
Key 6:   Manufacturing Value Analysis (Methods Improvement)
Key 7:   Zero Monitor Manufacturing
Key 8:   Coupled Manufacturing
Key 9:   Maintaining Equipment
Key 10:   Time Control and Commitment
Key 11:   Quality Assurance System
Key 12:   Developing Your Suppliers
Key 13:   Eliminating Waste (Treasure Map)
Key 14:   Empowering Workers to Make Improvements
Key 15:   Skill Versatility and Cross Training
Key 16:   Production Scheduling
Key 17:   Efficiency Control
Key 18:   Using Information Systems
Key 19:   Conserving Energy and Materials
Key 20:   Leading Technology and Site Technology

There are five levels of accomplishment for each key ranging from "not having a clue" to "world-class performer" with 1 being the worst rating and 5 being the best. The table below shows some sample definitions of the different levels.

Level	Description of Key Level
1	Little or no system in place. Reactive management. No continuous improvement.
2	System in place but needs improvement. Still reactive management. No continuous improvement.
3	Solid system in place that meets needs across most areas. Management more proactive than reactive. Continuous improvement efforts are taking place. Minimally acceptable system.
4	Systems meeting needs across all areas. Proactive management with well-structured continuous improvement program. Better than average.
5	World-class operation.

See the tables below for an example of how one company defines 2 of their 20 keys.

Level	KEY #15 – Skill Versatility and Cross Training
1	People do the job they are assigned and do not cross job lines.
2	A formal effort is under way to define skill versatility (SV) matrix for the area or department.
3	A SV matrix for all group members has been developed and is displayed. At least 1/2 of the group is trained on two other skills.
4	All group members can do at least three other jobs. Efforts are under way to extend the SV matrix to other work areas.
5	The SV matrix covers many areas. SV achievement is tied to compensation and promotional opportunities.

Level	KEY #17 – Preventive Maintenance
1	Operators do little or no preventive maintenance (PM).
2	An operator preventive maintenance (OPM) plan has been implemented for the most critical machines in the area.
3	An OPM plan has been implemented for all machines in the area and is generally followed by operators.
4	Operators begin to assume more complex PM responsibilities. An OPM action hardly ever is missed or late.
5	Operators truly own their machines and enthusiastically make every possible effort to maintain them in outstanding condition.

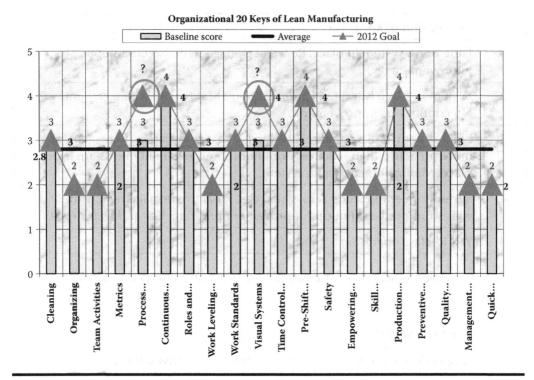

Figure 3.1

The 20 keys approach is very versatile in that within each category companies can develop their own definitions for each level of accomplishment. Organizations can also modify the key categories to describe and assess their organization better within the framework of its own industry and operations. After the initial assessment across all 20 keys, then the organization can choose the most critical keys to work on, in order to increase their ratings in that area. Once the areas of most immediate concern have been raised to an acceptable level then the next most crucial areas are addressed. Figure 3.1 is a sample 20 keys rating for a manufacturer in the medical device industry.

The 20 keys are just one approach to pursuing continuous improvement in your operations with an end goal of achieving world-class status and delivering delightful service. The most important thing is that the journey begin with an assessment and a "map" of how you plan to get to where you are headed.

Inverse Ishikawa (Fishbone) Diagram

The traditional use of an Ishikawa diagram is as a part of root cause analysis where the stem represents the problem being investigated and each branch

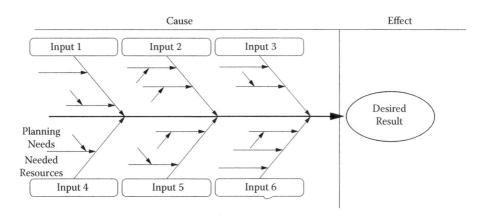

Figure 3.2

of the stem represents a category of potential causes. Unlike the traditional Ishikawa diagram where we are trying to determine the root cause of an adverse event that has already occurred, when we use the inverse Ishikawa diagram (IID), we are trying to determine what inputs are needed to bring about a desired future state (or event). In this case, the stem would represent the desired occurrence or output (if a process) and each branch would represent a category of inputs needed to bring about this desired future state. As shown in Figure 3.2, the first level subbranches of each branch are the needed resources and the second level subbranches represent the planning or action needs to acquire each resource.

When using the IID, there are two other tools that come in handy, brainstorming and affinity diagrams. This is how we populate the fishbone.

One common method for successfully doing brainstorming is to get a group of people together and go around the room with each person suggesting one idea at a time around a predetermined theme (problem, goal). They put each idea on a sticky note and place it on a large whiteboard. If a person has no idea to contribute, they pass on that round; if they have an idea later, they can voice it when it is their turn again. The brainstorming stops when everyone in the latest round has passed. It is important to note that all ideas are accepted at this point. None are evaluated or discouraged.

Once we have a list of ideas, a natural grouping of the ideas becomes apparent. The brainstorming team then begins grouping the sticky notes into four to six categories. If the team can't decide which group a particular idea belongs to, it is OK to duplicate the sticky note and put the idea in two separate groups. The groups are labeled and this finishes your affinity diagram. The labels are then placed on the branches of your IID. Next, the

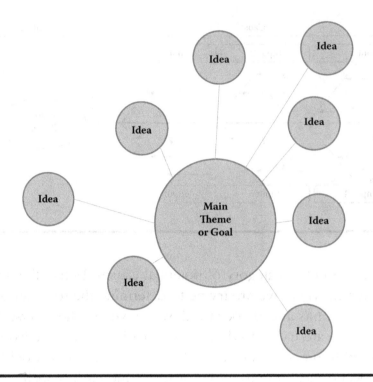

Figure 3.3

ideas from each group are placed on either a first-level subbranch or a second-level subbranch, based on whether it is a needed resource or planning/action need, respectively. During this sorting process, ideas are also evaluated. Some ideas are combined and those items deemed without merit by the group are discarded. These three tools when used in conjunction are very effective. But what do you do without a group to brainstorm with, when you are working by yourself? Then you can use a technique called clustering.

With clustering, you draw a circle in the middle of a piece of paper and write your goal, main theme, and so on in the middle. Then as quickly as you can, write down around the circle, words, phrases, ideas, or associations that come to mind. Once you have written down all of your ideas, then circle each and draw a line from each circle to the center circle. When you are done, it should look something like Figure 3.3. It is OK to take a break, then come back to see if you can add more. It is also OK to add more circles if you think of them later when you are doing something else. The goal of this exercise is to take advantage of the creative ability of the subconscious mind, and it really works. It is also great for writers who have writer's block, as the start of a free writing exercise. From here you can now

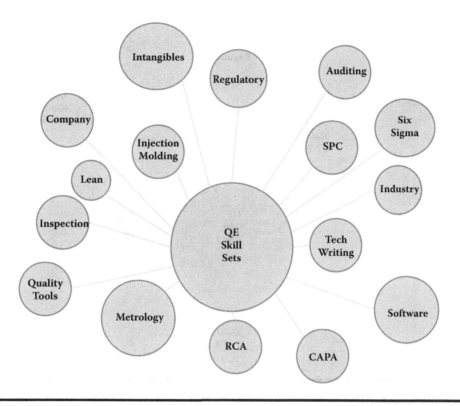

Figure 3.4

create your own affinity diagram and begin populating your IID. If this is a little abstract for you, let's take a look at a process that I went through myself several years ago.

It had been a little over a decade since my last position in quality as a quality manager in a different industry. My most recent employer had gone out of business during the recession of 2009 and I applied for and took a position as an inspector to get my foot in the door with a local manufacturer. I was both new to the company and new to the industry. I set two goals for myself, one professional and one financial. The professional goal was to leverage my past experience and education to be promoted to quality engineer within two years. The financial goal was a little more daunting: increase my salary by an average of $10,000 per year for the next three years. I used a combination of the ASQ (American Society for Quality) certified quality engineer body of knowledge and the job description for quality engineer at the company where I was to come up with the skill sets that I needed to acquire to become a quality engineer, and created my cluster diagram. Initially before editing, my cluster diagram looked like Figure 3.4.

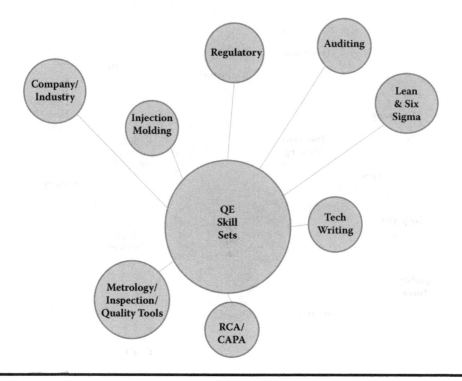

Figure 3.5

Once I had combined, discarded, and changed some of the categories I was left with the simplified diagram shown in Figure 3.5.

I then ranked them according to the classifications below which allowed me to set priorities, as I populated my fishbone in Excel, as seen in Figure 3.6. My goal was to get every classification to 3 or higher with at least half of them 4 or higher, and then apply for the next available opening. I allowed myself 18 months to accomplish these goals plus another 6 months (total 24) to attain the promotion:

5. Extremely knowledgeable or skilled; could train others
4. Very strong experience or skill
3. Adequate experience or skill
2. Some experience or skill, need improvement
1. Very little experience or skill

I devised the list of resources and actions below to place on the branches of my IID. Also see Figure 3.7. I set priorities, developed a timeline, and aggressively pursued my goals, checking progress against milestones every quarter.

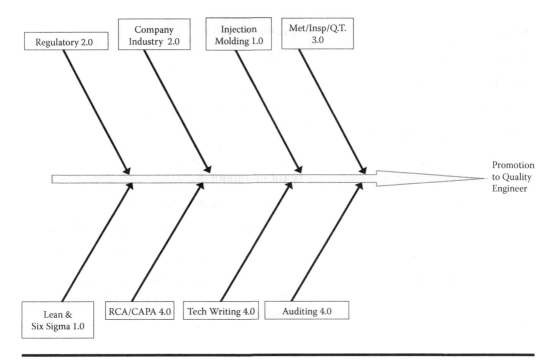

Figure 3.6

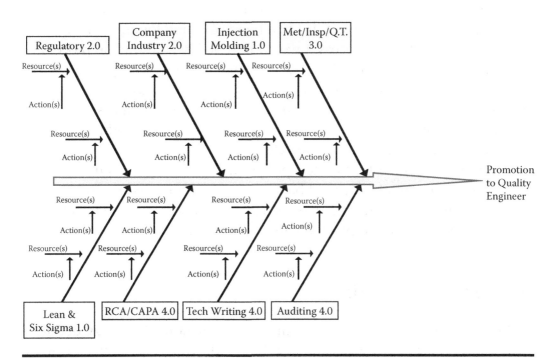

Figure 3.7

Resources	Actions Required
College courses	None
Mentors: external and internal	Seek out mentor(s)
Self-study	Request permission from supervisor and instructor to attend internal training
Formal (classroom) internal training	
Formal external training	Register for class and seek tuition reimbursement
Certification	Attend professional conferences
On-the-job training	Identify and pursue professional certification
Previous experience	Request permission to attend external training
Previous education	Read industry periodicals, training manuals, textbooks, and other opportunities for self-study

Results

I achieved promotion in 14 months and financial goals in 34 months.

This same process could have easily been started with a customer need or internal improvement project. If I had been planning for an improvement to a manufacturing process, I would have followed the same process with only minor exceptions. The clustering, filtering, and sorting would have been accomplished in the same manner. One difference would be that I would use the 6Ms—man, machine, method, material, measurement, and mother nature—for the initial categories on my IID, then add or delete categories based on the results of my clustering. Next I would develop a plan with timeline, milestones, checkpoints, and feedback. Then finally, after "planning the work" it would be time to roll up our sleeves and "work the plan." Thus you have two key elements of success when deploying the Kano model: extensive planning and hard work.

Hoshin Kanri X-Matrix

Hoshin Kanri is a strategic planning and management methodology developed to capture strategic goals and distill them down to measurable objectives and targeted action items. It starts with the organization making up to four vision statements of where the company wants to be in three to five years. *Hoshin* means "compass needle" or "direction" and *Kanri* means "management" or "control." The X-matrix simply denotes a common format for presenting goals and objectives. For maximum effectiveness, the planning team must be both cross-functional and from the tiers of management high enough to affect strategic goals.

There are seven steps to the Hoshin Kanri planning process:

1. Identify key business issues facing the organization.
2. Establish measurable organizational objectives that address these issues.
3. Define the overall vision and goals.
4. Develop supporting strategies to achieve these goals.
5. Determine the tactics and objectives that facilitate each strategy.
6. Implement performance measures for each key organizational process.
7. Measure fundamentals of organizational success.

These steps are captured on X-matrices, each tied to different levels of the organization and to each other. The process starts with identification of the mission statements, and determination of strategies for how to flow the mission information and implementation from executive- to ground-level employees. Figure 3.8 is an example of a high-level X-matrix. The table is meant to be read starting at the darkest tone "six o'clock" position, then moving clockwise.

The next level of the organization, usually middle management, then uses its own X-matrix as a guideline for implementation of the corporate strategies as shown in Figure 3.9. And finally to ensure consistency of effort and desired results the first-level manager/supervisors and their employees get their own X-matrix to implement. Please see Figure 3.10.

Action plans are developed for each X-matrix level to ensure consistency of purpose and quantifiable results. An example of an action plan is seen in Figure 3.11.

In more formal (read complicated) implementation of Hoshin Kanri the plan is monitored and assessed using previously identified metrics and reported as tables. The four Hoshin table types are:

■ Hoshin review table: During reviews, plans are presented in the form of standardized Hoshin review tables, each of which shows a single objective and its supporting strategies.

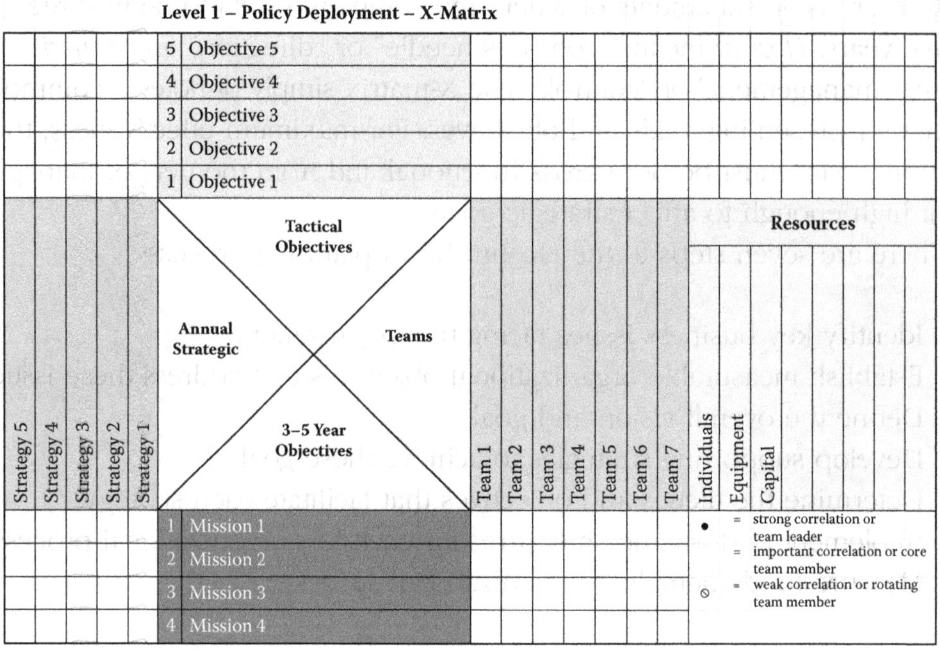

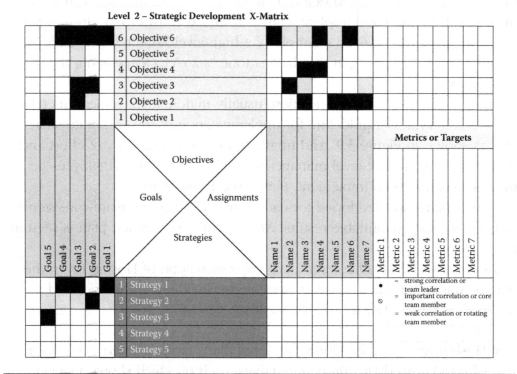

Figure 3.8

Figure 3.9

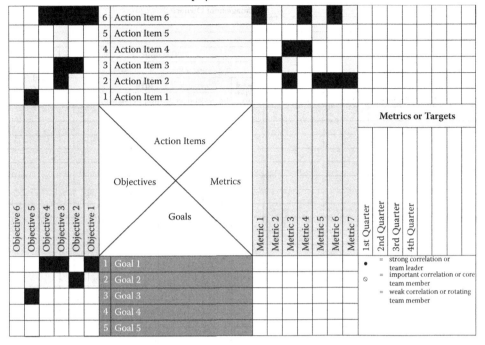

Figure 3.10

- Strategy implementation table: Implementation plans are used to identify the tactical objectives or action plans needed to accomplish each strategy.
- Business fundamentals table (BFT): Business fundamentals are the core elements that define the success of a key business process, and are monitored through its related metrics. Examples of business fundamentals are safety, people, quality, or cost.
- Annual planning table (APT): Stepped down from the X-matrix showing a three- to five-year strategic outlook, the APT records the organization's objectives and strategies in the annual planning table. The APT is then passed down to the next organizational structure.

Metrics are charted and reported on a monthly basis to ensure that the plan remains on track and that no adjustments need to be made. As the Hoshin planning process is meant to be cyclical, with the plan from one year forming the basis of the plan for the next, an overall review of the plan is required to be done annually. It is important to remember that the annual assessment is not just against planned activities and results, but also against your "organizational compass." Have the customers' current and perceived future needs remained the same since the last assessment?

Action Plan																		
Improvement Priority:				**Management Owner:**													**Date Created:**	
Team:																	**Next Review:**	
Background:																		
Relation to Annual Objective:				Timeline													Status Red, Yellow, Green	
					= Original Plan		x	= Complete										
				Planned Complete Date	2014													
Action Step/Kaizen Events	**Owner**	**Deliverable**			Jan	Feb	Mar	Apr	May	Jun	Jul	Aug	Sep	Oct	Nov	Dec		

Figure 3.11

Quality Function Deployment (QFD)

Quality function deployment is a process that uses a planning matrix to provide a direct link from customer needs and wants to product or service requirements. QFD is also referred to as the "house of quality" and sometimes as "listening to the voice of the customer." Actually, QFD would be better described as a series of linked processes (or houses):

- The first house translates customer needs and wants (VOC) into functional requirements.
- The second house translates functional requirements into product design requirements.
- The third house translates product design requirements into process design requirements.
- The fourth house translates process design requirements into the production realities of process control.

The steps in the QFD process are:

1. Plan. Determine objectives and data needed.
2. Collect data and begin to implement plan.
3. Analyze and understand data using the QFD matrix to identify the opportunities for improvement and innovation.
4. Utilize the knowledge gained to improve designs and achieve customer value.
5. Evaluate and improve the QFD process (continuous feedback loop).

So how do we populate this matrix? We use the technical specifications as well as information from the customer on end function as well as usage in the next process step. We also use information from surveys, direct observation, customer experience, quality records, focus groups, and joint development projects.

Wants should be quantified and weighted by importance to prioritize and ensure that the most important criteria are defined first. Customer process characteristics describe how things are done and should specify design requirements, operational factors, and human factors. There is no single accepted format for a QFD matrix, however, there are certain consistencies. You construct from left to right, with customer needs on the left, relationships in the middle, and customer perceptions to the right. Critical to quality (CTQ) characteristics and correlations form the top horizontal portion of the matrix. The technical importance rating forms the base of the "house." One example of a QFD matrix template is seen in Figure 3.12.

The strength of the relationships between customer functional needs and organizational requirements should be established and planned for. Icons are used in the QFD matrix to represent the strength of the correlation between individual inputs (needs, wants, requirements) and individual outputs (specifications, results, requirements). The filled circle indicates the strongest relationship; the open circle, a moderate relationship; and the triangle a weak relationship. It should also be determined if the correlation between an input and output is positive or negative. A negative correlation is indicated by an "x" and a strong negative correlation is indicated by a "#". See Figure 3.13.

Now let's take a look at how we would populate a QFD matrix based on the scenario from earlier in this book, where a gentleman wants to purchase

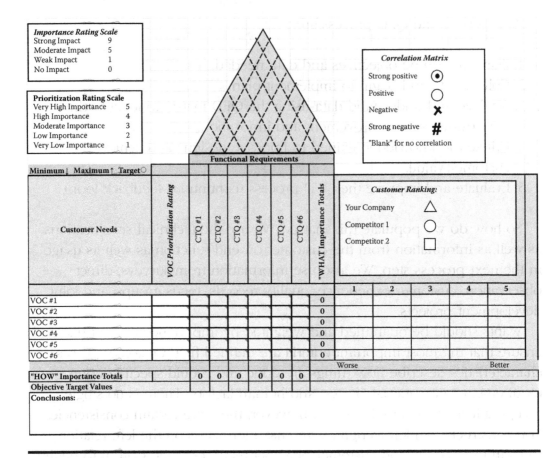

Figure 3.12 (From Munro, R.A. et al. *The Certified Six Sigma Green Belt Handbook.* ©2008 ASQ. With permission.)

a walking cane. You can see that the matrix may start to become involved, even with a simple scenario such as this one. Filling out a QFD matrix takes time, discipline, and work. However, having said that, the QFD matrix is an incredibly powerful tool that, when filled out properly, can provide information that will allow us to plan, identify needed controls, and allocate resources. As I learned in Six Sigma Black Belt training 80% of a successful project is time spent during planning.

Directions

1. Enter customer needs into customer needs column in the voice of the customer (VOC) boxes.
2. Enter functional requirements across the top in the critical to quality (CTQ) boxes.

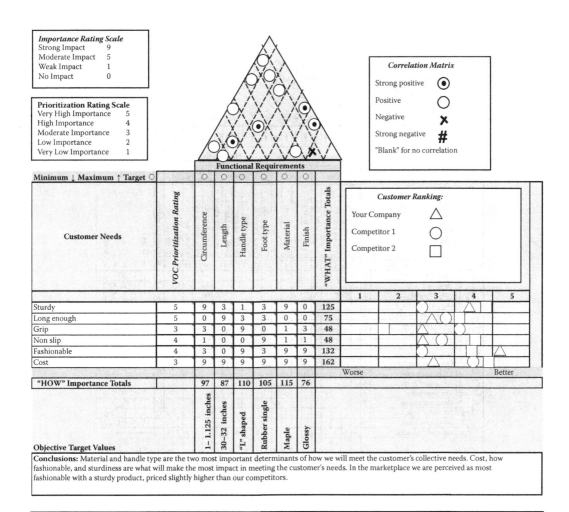

Figure 3.13   **(From Munro, R.A. et al. *The Certified Six Sigma Green Belt Handbook.* ©2008 ASQ. With permission.)**

3. Indicate whether each requirement is a minimum, maximum, or target value.
4. Rank prioritization for each need from one to five, with five being highest priority.
5. Use externally derived data to rank your company and your primary competitors on delivering key customer needs, and place symbols accordingly.
6. Write in actual target values for each of the CTQ across the bottom of the matrix.
7. Once the spreadsheet has completed its calculations, type in what conclusions you are able to make in the "conclusions" box.
8. Take action.

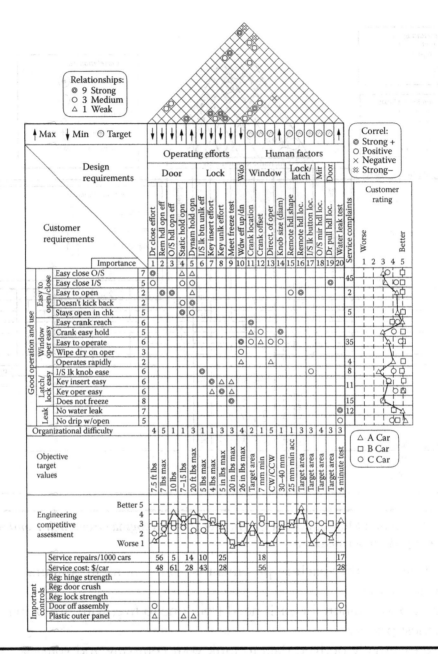

Figure 3.14 (From Munro, R.A. et al. *The Certified Six Sigma Green Belt Handbook.* ©2008 ASQ. With permission.)

Thus, the QFD matrix provides a concrete means of assessing how close we are coming to providing delightful service, by taking into account feedback from the field as well as planning and execution, all within one model. Another example of a completed QFD matrix is seen in Figure 3.14. In this case the matrix is looking at a much more complex manufacturing scenario.

Chapter 4

Deployment Tools

As stated previously in this book, organizational discovery and delivery of delightful service is contingent upon extensive planning (and patience) as well as implementation, monitoring, feedback, and adjustment as necessary (discipline and hard work), all existing within an environment of continuous improvement. In this chapter, we discuss the tools used in the project (or business/nonprofit launch) implementation, monitoring, and feedback as listed below.

Implementation

- Supplier–Input–Process–Output–Customer (SIPOC) diagram
- Value Stream Maps
- Gantt Charts
- Critical Path Method (CPM)

Monitoring and Feedback

- Surveys
- Control Charts
- Audits
- Gemba Walks

Audits

So what do audits have to do with customer service of any kind, you might ask? Actually, the audit function has both direct and indirect ties to the Kano model specifically and customer service applications in general. The audit itself is a service, and when used to verify compliance as it most often is, then verification of compliance to regulatory and customer specifications directly ties back to supporting customer satisfaction. Going beyond compliance auditing, let's look at one common definition of an audit as an assessment against criteria or requirements. Criteria can be attainment of project milestones, improvement initiative results, keeping up with a timeline, and so on.

Reference standards are external documents such as regulations, contracts and International Organization for Standardization (ISO) standards that establish minimum requirements, in other words, explicit customer needs and expectations. *Performance standards* are internal documents such as standard operating procedures (SOPs), work instructions, and drawings and other similar documents that describe how requirements will be met and against which personnel performance must be audited.

When conducting an audit, the auditor should first assess the company documentation against the related reference documents. Any findings would then be against company documentation or the quality management system. Then the auditor should match employee actions and records (performance) against what is stated in her own internal documentation. Any findings noted would be against performance of actions as required. A memory aid and visual depiction of this process is seen in the "W Factor" illustration in Figure 4.1, along with a sample checklist template in Figure 4.2.

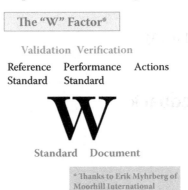

Figure 4.1

In other words, if during the documentation review, a performance standard is found to be missing or in violation of a reference standard, then a noncompliance is written against the reference standard (left "valley" of the W). If observed actions aren't as documented, then a noncompliance is written against the internal document or performance standard (shown in the right "valley" of the W). This methodology avoids the confusion of siting multiple sources when referencing a single noncompliance. By having a structured internal audit program with trained auditors and regularly scheduled audits, an organization can verify that they are working in a manner that will produce a product or service that meets customer needs and expectations. Audits can also identify risks to safety or disruption in operations as well as serve as a catalyst for developing an environment of continuous improvement. These are all critical elements to an organization wanting to deliver delightful service to its customers.

Sample Checklist

Figure 4.2 shows a sample audit checklist.

Audit Questionnaire – Organization Improvement Project Q1 2013			
Requirement/ Goal	Question:	Objective Evidence:	Comments (Indicate Major, Minor or Opportunity)
	☐ Y ☐ N ☐ N/A		
	☐ Y ☐ N ☐ N/A		
	☐ Y ☐ N ☐ N/A		
	General Observations ☐ Y ☐ N ☐ N/A		

Figure 4.2

	Jan 19	31	Feb 16	28	Mar 16	31	Apr 13	27	May 18	31	Jun 15	29	Jul 13	27	Aug 17	31
Define																
Measure																
Analyze																
Improve																
Control																

Define –	Confirm problem, establish project charter, determine resources, develop plan, design of experiments (DOE)
Measure –	Measurement system evaluation, take measurements
Analyze –	Analyze data, confirm hypothesis, modify plan as necessary
Improve –	Implement improvement and verify that it works
Control –	Standardize improvements and monitor for three months

Figure 4.3

Gantt Charts and Critical Path Method (CPM)

Gantt Charts

Gantt charts are horizontal bar charts used to track project timelines. The project aspects being tracked are typically placed on the *y*-axis and time is tracked on the *x*-axis. In preparation for creating your Gantt chart identify critical project dates, project steps, and how long it will take to complete them. Place the required completion date at the end of the timeline and begin to work backwards. I usually like to allow myself at least a 10% window on the timeline (if due in 10 days, set project completion at 9 days on the timeline; if there is a 10-week due date, then project completion is set at 9 weeks; and so on) to allow for any unforeseen delays. For aspects of the project having flexibility as to when they must be completed, plan for them to be done later in the project lifecycle. That way, if another aspect of the project is delayed, one of the project components having flexibility as to when they had to be completed, can be pulled forward so that resources don't sit idle, and there is less likelihood of disrupting the overall project timeline. Figure 4.3 shows a Gantt chart taken from a recent Black Belt project on which I was working. On the *y*-axis you will see the different phases of DMAIC,* but it could

---

* DMAIC is the acronym for the Define-Measure-Analyze-Improve-Control methodology most often followed in Six Sigma improvement projects.

have been project phases as identified in a project plan, or another type of milestone. The *x*-axis reflects dates.

Critical Path Method (CPM)

Another important aspect of creating an effective Gantt chart is setting priorities. This is where the critical path method (CPM) comes in. On a project with multiple pathways to completion, the critical path is the path from start to finish that requires the most time. In a project with only one pathway to completion, that path is by default the critical path. The only way to shorten the lifecycle of a project is by positively affecting the critical path. Thus, by identifying the critical path, project members know where to concentrate resources if the completion time of the project is a concern. Sometimes sequencing can allow us to shorten the critical path. A CPM diagram is shown in Figure 4.4.

The four potential pathways are ACEFGI, ACEFHI, ABDFHI, and ABDFGI. The critical path is ACEFHI which would take 45 days. The best practice would be to determine the critical path, optimize sequencing of operations, and then create a Gantt chart. For a single path project, those steps that are sequence optional should be pushed as close to the final phases of the project as possible, and any portions of the project that can be accomplished simultaneously should be scheduled accordingly. Finally, the feasibility of procuring additional resources to assist if needed, with the most time-consuming aspects of the project, to speed up the process should be explored during the project planning phase.

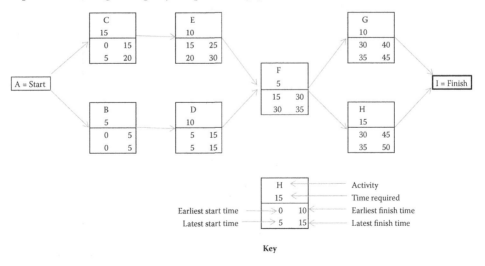

Figure 4.4

Supplier–Input–Process–Output–Customer (SIPOC) Diagram

SIPOC diagrams are a tool used to identify all of the high-level aspects of a project that need to be accounted for—suppliers, inputs, the process, outputs, and customers—before work can begin. A SIPOC diagram can be reflective of either a single process or series of related processes. Creation of a SIPOC diagram is usually followed by development of a process map. In that case, the inputs and output for the SIPOC and its related process map should match identically.

The SIPOC diagram plays a role in further developing the Kano model by identifying for an organization, the customer, their needs, and the resources needed to supply those needs. It also clearly depicts which outputs should go to which customers, in the case where there might be multiple customers. The SIPOC diagram is a "living" document that should be updated as required throughout your project. Figure 4.5 is a common SIPOC template format. SIPOC diagrams will not always include the "Requirements" columns, although ours do.

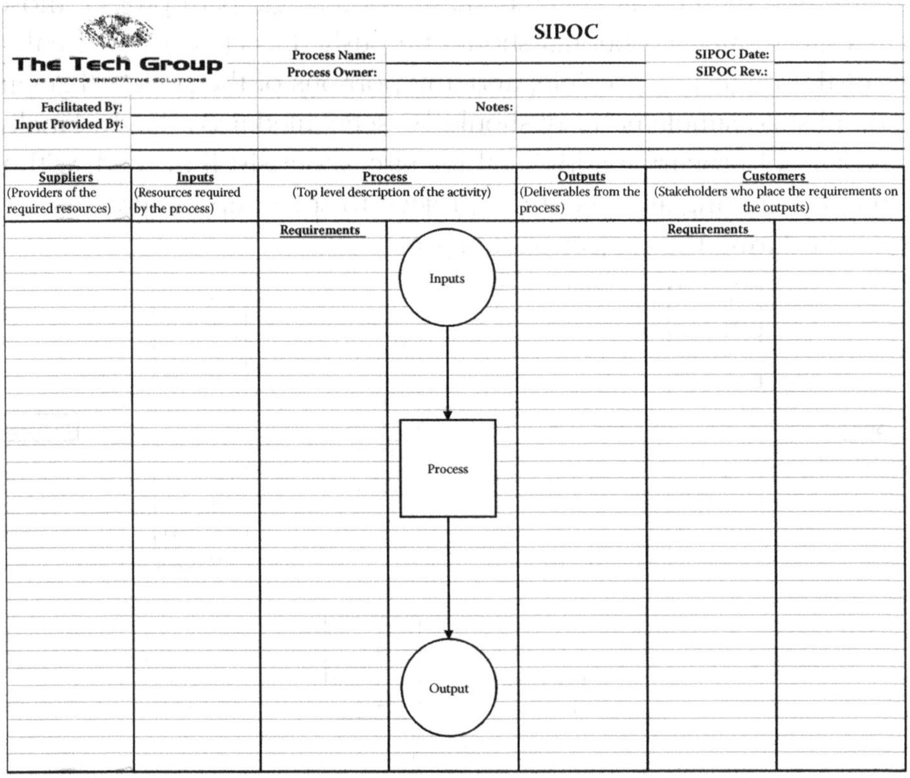

Figure 4.5

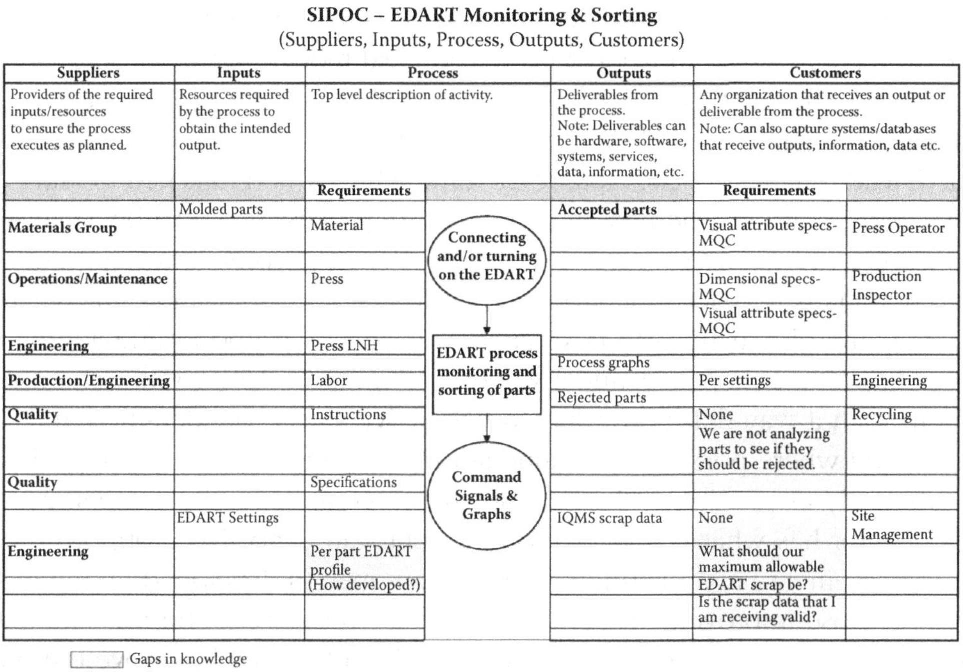

SIPOC – EDART Monitoring & Sorting
(Suppliers, Inputs, Process, Outputs, Customers)

Suppliers	Inputs-Requirements	Inputs	Process	Outputs	Outputs-Requirements	Customers
Providers of the required inputs/resources to ensure the process executes as planned.	Resources required by the process to obtain the intended output.		Top level description of activity.	Deliverables from the process. Note: Deliverables can be hardware, software, systems, services, data, information, etc.	Any organization that receives an output or deliverable from the process. Note: Can also capture systems/databases that receive outputs, information, data etc.	
		Requirements		**Requirements**		
	Molded parts			Accepted parts		
Materials Group		Material	Connecting and/or turning on the EDART		Visual attribute specs-MQC	Press Operator
Operations/Maintenance		Press			Dimensional specs-MQC	Production Inspector
					Visual attribute specs-MQC	
Engineering		Press LNH	EDART process monitoring and sorting of parts	Process graphs		
Production/Engineering		Labor			Per settings	Engineering
Quality		Instructions		Rejected parts	None	Recycling
					We are not analyzing parts to see if they should be rejected.	
Quality		Specifications	Command Signals & Graphs			
	EDART Settings			IQMS scrap data	None	Site Management
Engineering		Per part EDART profile (How developed?)			What should our maximum allowable EDART scrap be?	
					Is the scrap data that I am receiving valid?	

☐ Gaps in knowledge

Figure 4.6

Figure 4.6 shows the SIPOC from the Black Belt project that I mentioned previously, pertaining to an electronic process monitoring system. Gaps in knowledge are shaded in gray tone and should be filled in as the project progresses, additional knowledge is acquired, and the SIPOC diagram is updated.

Value Stream Mapping

To understand value stream mapping, you must understand what is meant by value. For an operation to have value, three conditions must be in effect. A transformation must take place between process inputs and process outputs, the customer must be willing to pay for it, and it must be done right the first time. The value stream is that collection of activities, both value- and non-value-adding, as well as receipt of materials and components from suppliers and delivery to the customer, that makes up an organization's operation.

Three other terms to be aware of are *cycle time*, *lead time*, and *takt time*. Cycle time is the average time for a product to go through a process step. Lead time is the total of all cycle times plus waiting times (waste) from receipt of the order to delivery at the customer that it takes to produce and deliver one unit of what has been ordered. Takt time is the consumer consumption rate or demand for an individual item. The ideal model matches lead time to takt time, so that as a unit of product or service is consumed, another is immediately delivered. It should be noted that as waste in the system is eliminated, lead time approaches total cycle time which when matched to takt time, leads to an environment where optimum process flow can be achieved.

A value stream map is most often used as an enterprise-level process map that takes into account suppliers, customers, transportation, intelligence transfer, and obstructions to flow, as well as value- and non-value-adding processes. The goal in creating a value stream map is to view the organization in such a way that non-value-adding activities can be identified and eliminated where possible and value-adding activities can be accomplished as efficiently as possible. When this has been completed, the value stream map creator has what is considered a current state map. Once opportunities for improvement have been identified, a new map is developed taking into account those improvements, and that map is considered a future state map. Then improvement projects are planned and implemented to move the organization from current to future state.

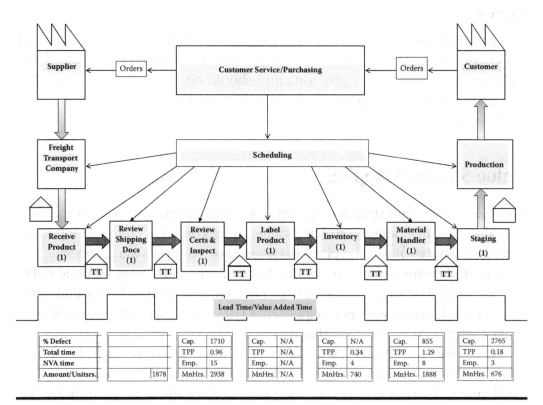

Figure 4.7

A simple value stream map can be seen in Figure 4.7. Information transfer is shown at the top of the map, moving from right to left, and material and product transfer are shown at the bottom, moving from left to right. The two "columns" show transport of materials or product from suppliers (left) and to customers (right).

Some purists prefer to hand-write their value stream maps, taking prompts from the team, crafting the map, and filling in details as the organization's processes are described, rather than relying on software. Figure 4.8 is a picture of a value stream map hand drawn on a whiteboard.

Although by definition, a value stream map is an enterprise-level look at operations, value stream mapping methodology can be used to look at specific sections of an organization as well. For example, value stream mapping can also be incorporated into the internal quality audit program for a more robust and versatile process. During a traditional audit, the following types of questions might be asked.

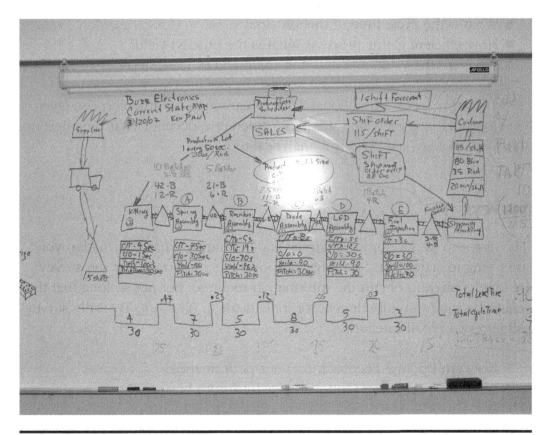

Figure 4.8   (Thanks to Chris Hayes, CEO of Impact Performance Solutions, for the use of this image.)

Compliance-Based Questions

- What is the requirement?
- What do our documents say?
- Do we do what we said we would do?
- Are our records complete and consistent?

During a value-add or continuous improvement audit the additional questions below might also be added to the checklist. By incorporating value-add questions in an audit, organizations can not only confirm compliance to regulations and company procedures but also identify opportunities for improvement in a way not done previously.

Value-Add Questions

- What is the value proposition of this process?
- What is the cycle time of this process?
- How efficient is this process? What is the process yield?
- What are the value-add and non-value-add process steps?

So whether using a traditional value stream map for an enterprise-level view or looking at the value stream of different aspects of your organization, this methodology is a powerful analytical tool.

Surveys

The first step in developing a survey is determining what information you are trying to acquire and for what purpose. It is important that the survey be long enough to obtain the information required, but not so long that the length becomes an impediment for individuals to fill it out. Typically surveys fall into two broad categories:

1. How am I doing? Feedback on your performance
2. What are you doing? Opportunities for future business

It is important to try to determine how important the product or service is that you provide to your customer in addition to how good a job you are doing in providing it. A simple way of thinking about this relationship is the statement below.

$$\textbf{Q}uality + \textbf{I}mportance = \textbf{V}alue$$

It is also extremely important to convey to the survey recipients that they are doing something important and worthwhile by completing your survey. Some examples of important information to find out about your customers are shown below. You might not ask all these questions for every survey, but this type of information is important to have:

■ How am I doing?
■ What can I do better?
■ What technologies can I invest in to better meet future needs?
■ What opportunities might we have with you in the future?
■ How do we compare with our competitors?
■ How do we compare with our own past performance?
■ Would you recommend us to someone else?
■ How loyal are you to our brand?
■ What attracted you to us initially?
■ Why do you stay with us?

Some questions are specific to whether your organization has delivered on its promise. A customer might stay with you for reasons other than why you were initially chosen, and so it is important to draw a distinction between these two questions. This information can lead to better marketing efforts as it may point to features of your product or service that are valuable to the customer that may not have been emphasized in previous marketing initiatives. It will also allow your organization to shore up weaknesses and build on strengths, as perceived by your customer.

One other important thing to consider when conducting surveys is when and how often to send them out. There is so much "noise" in our everyday lives right now that people become annoyed by interruptions such as television commercials, computer pop-up ads, or surveys that are inserted into their lives. With that being the case, a repeat survey should be sent out at the absolute minimum frequency that is needed to acquire the information that you need when you need it. Similarly a "one-time" survey should be sent out far enough in advance that the information gathered can be analyzed and used for the purpose for which it was intended. Ideally the survey should not be conducted during the busier times during the day or year for respondents (i.e., the holidays or tax time, if mailed, or dinner time if calling). Also be sure to let the survey respondent know what the information will be used for, and provide assurances that the information and contact information will be put to no other use.

Finally, make the survey as easy to answer as possible. Use a ranking scale of either 1 to 5, or 1 to 10, with clear definitions for each ranking, or multiple choice questions whenever possible. This strategy will also make it easier to compare survey results from different groups. Don't ask too many open-ended questions, because then your survey recipient may begin to feel as if he is writing an essay, not the effect that you want. A well-crafted survey is a vital tool in the effort to capture the voice of your customer and craft your products or services to provide the most value from the customer's perspective, and thereby deliver truly delightful service.

Control Charts

Control charts are used to plot data that conform to a normal (or bell-shaped) distribution pattern. Learning to interpret control charts is beyond the scope of this chapter. It is my goal to share with my readers the benefits of having control charts and how they can be used once implemented. The two most important pieces of data for describing a population are the location (where centered) and dispersion (how spread out). The three primary purposes for use of control charts are to monitor the process, identify special and common cause variation, and show if the process is in control. This can be a manufacturing process where a product is being made or a transactional process such as loans made or documents typed.

The power of control charts comes in their predictive ability. Much of the data that we review and analyze relating to our processes have a population that falls into a normal distribution pattern. The nature of this pattern allows for predictive modeling which we use in design of experiment for validation and optimization of processes and also when determining if a process is capable or in control. In addition, control charts allow us to identify special and common cause variation in a process.

Common cause variation is variation that is normal to the process. Attempts to adjust a process for common cause variation often make things worse. *Special cause* variation is variation that is (typically) outside the three Sigma control limits caused by an external factor. Root cause should be determined and corrective action applied to eliminate special cause variation.

There are two broad categories of control charts: variable and attribute. *Variable data* are continuous and usually describe a measurement of some kind such as 2.75 pounds or 18.13 inches. *Attribute data* are discrete and typically describe a state (up/down, on/off, 1/0, blemished/not blemished, etc.) or a count such as 16 chickens or eight errors per document.

Some most commonly used attribute charts are

- *P-chart:* Percentage defective
- *NP-chart:* Number of parts defective
- *U-chart:* Defects per unit
- *C-chart:* Defects within the sample

Some most commonly used variable charts are:

- *Individual moving range (IMR) charts:* Individual data points
- *X-bar/R charts:* Subgroups

The most important thing in using control charts is spotting trends that will allow an organization to make corrections for negative trends before they become a problem that could cost money or, worse, customer dissatisfaction. They also allow for reinforcement of actions that may be (sometimes unknowingly) causing positive trends. This is accomplished by forming hypotheses related to the process outputs, based on changes to the inputs and watching the data to see if you are correct. Bottom line: with whatever you are doing, whether making a product or delivering a service, who wouldn't want to be able to predict the future to some extent?

Gemba Walks

Sometimes called "management by walking around," the term *Gemba walks* comes from the Japanese word *gemba* meaning "to go where the action is," where value is created for the customer. Going to gemba is not just a course of action, it is also a philosophy espousing that top management get out of their "ivory towers" and go out on the ground floor to interact with front-line employees/volunteers as one means of finding out what is going well and what needs fixing in their organization.

Although Gemba walks aren't scripted and don't use checklists, there should be an overall theme for each walk such as Lean implementation, performance, safety, and so forth. However, that's not to say that those doing the walk shouldn't be open to hearing concerns of any type if broached by employees. Some benefits to be derived from Gemba walks are

- Improved relationships within the organization
- Increased employee morale
- Increased knowledge base for upper management
- Positive visibility of upper management
- Continuous improvement

Gemba walks are spoken of and are meant to originate with the highest level of management but should also be used by each subsequent lower level of management with the same potential benefits.

Case-In-Point: Power of Gemba Walks

On a recent trip to South Africa, I heard the best story on the power of Gemba walks which I now share with you. The names and some facts have been changed to protect both the innocent and the guilty. However, the truth at the core of the story remains and is evident in the telling.

There was a senior vice president (SVP) at a multinational corporation who was sent to run a new site and charged with improving productivity and reducing waste. As one means to achieve that goal she instituted the practice of Gemba walks, going where the action was. At that time, it was practice at the site for all the directors to meet monthly to go over performance against annual goals. It was thought by some that the director for logistics and planning might be aggressively optimistic, let us say, in presenting his numbers. After all, how could the numbers be accurate and the site having the troubles they were having (among others) meeting their on-time-delivery goals.

So the SVP decided to take one of her first Gemba walks down to the warehouse which is where she met Jacques, the warehouse supervisor. Jacques reported to the materials manager, who reported to the director of logistics and planning. He is the kind of employee that every organization has or wished they had. At this time, Jacques had been with the company for 15 years, rarely missed a day of work, knew the warehouse inside and out, and was passionate about his job. The SVP introduced herself and explained that she was touring the plant to become acquainted with the personnel and how things are done. The two exchanged pleasantries for a few minutes with Jacques explaining how he had come to work for the company for 15 years and how much he enjoyed his job and working for the company.

Everything was very positive right up until the SVP asked, "So how do you like our inventory system?" Then Jacques replied, "It's terrible!" "What do you mean?" asked the SVP.

> Well we often order too much of some things and not enough of others. Do you see these pallets over here? We rarely use these items and this pallet has been here for almost two years. That pallet there on that top shelf hasn't been moved since our physical inventory last year. Just look around, and you can tell by the amount of dust on certain items that we ordered too much of it, and that it's been here way too long. I've mentioned these things before, but nothing was ever done.

When Jacques stopped to take a breath, the SVP replied, "This definitely needs to be improved upon, but are we ordering the right amounts of anything?" Jacques said, "Sure, this pallet right here usually turns in about four days and that pallet by my office, every three to five days depending on the time of the year. I would say that probably about half what's here is in adequate quantity." "Well Jacques, you've given me a lot to think about," the SVP said. "Thanks very much for your time."

Indeed the SVP did do a lot of reflecting on what she saw and was prepared for the next plant operations meeting which took place during the following week. When the director of logistics and planning got up to give his report, he stated that he was proud to report that the department had achieved its goal of having an average inventory turn of fewer than four days. The SVP asked "Have you confirmed those numbers?" "Yes, we have double-checked all of our numbers." "Those numbers reflect an improvement that sounds almost too good to be true this quickly. Are you sure they are correct?" asked the SVP. "Absolutely," replied the director of logistics and planning. "I double-checked the numbers myself. We have without question achieved an average inventory turn of 3.9 days." "Well that's not what Jacques says." The director of logistics and planning then finally showed his "true hand" with this response …

"Who is Jacques???"

The SVP successfully went on to turn the site around through accurate analysis of and appropriately responding to metrics, by encouraging "ground up" improvement programs and by continuing to "go where the action is." The director of logistics and planning as well as the other site directors, now seeing the benefit of Gemba walks, joined the SVP in this practice with outstanding results.

Chapter 5

Metrics and Monitoring

Why Bother?

A desire without specifics is just a wish. Specificity leads to goal setting. To achieve long-term strategic goals, there has to be a series of incremental tactical objectives that will span the path from current to desired future state. These objectives, for them to be useful, must be measureable, and that's where metrics come in. Metrics provide an objective and universally understood means of determining if an objective or goal has been met. According to Eric Ries in *The Lean Startup*,* metrics should be "actionable, accessible, and auditable." Actionable so that you can do something to affect them. Accessible so that they make sense to the users. Auditable so that they can be monitored and tracked. But before we go any further, we have to talk about process monitoring.

Organizations have two broad categories of processes, both equally important: core processes and support processes. *Core processes* are those that create or develop the product or service to be delivered to the customer. *Support processes* are those processes that need to be there in order for the other processes to function. Both need to be monitored, but why? We monitor processes to know if we are receiving the expected or desired results based on our inputs to those processes. We also monitor processes in order to make predictions based on trends in data. This allows us to know when

---

* Ries, E. 2011.

we need to make adjustments to prevent a negative occurrence, or when to reinforce actions that have yielded positive results, or when to do nothing at all. Generally a process should be monitored at its beginning, middle, and end. Monitoring at the beginning tells us if our process inputs are acceptable and if beginning conditions are as expected. Monitoring in the middle tells us if our process is operating as expected, and monitoring at the end tells us if our process outputs are as expected given the process inputs; it also tells us if all of the critical inputs to a process have been identified. This is true of a single process (department or function) or a group of interrelated processes (system, business, or organization).

So how is this monitoring accomplished? Process monitoring can be accomplished using auditing, via metrics, or both.[*] Before we get to metrics though, we have to talk about their forerunner, measurable objectives. The relationship between measurable objectives and metrics is simple. Measurable objectives set a target or desired state. Metrics are the measurement that tells us how on target we are. For example, "a speed limit is a measurable objective (maximum limit) and the car speedometer (or police radar) provides the associated metric."[†] To summarize the preceding paragraphs, there should be a flow down from strategic goals to tactical (and measurable) objectives that are monitored using established metrics.

In addition to what metric is taken, it is also important as to where in the process it is taken, how often is it taken, and its relationship to other metrics being used. When choosing metrics one must recognize that in our current information society, in business as in life, we collect much more information than we will ever use. Data must be analyzed for them to become information. The trick is to decide which data are important to monitor and analyze if we want to achieve our objectives. Also, how is that information related to organizational goals and interrelated with other information? To get the most out of metrics, we must first understand that they can be categorized in different ways. For the purposes of this book, we examine two ways to categorize metrics as seen below:

By Level

- Strategic
- Operational

---

[*] Okes, D., 2013.
[†] Okes, D., 2013, p. 19.

By Type

- Core Process or Support Process
- Internal or External
- Quantitative or Qualitative

Strategic-level metrics typically reflect organizational placement within their market and might be financial, market share, or market surveys (providing customer feedback or existing perceptions in the marketplace about the organization, its products or its services). These also include metrics tied to measurable objectives that would drive the organization toward its strategic-level goals. Examples of supporting metrics at a strategic level might be the number of new products introduced per year or production (or service delivery) capacity.

Operational-level metrics have to do with how effectively and efficiently the organization is performing on a day-to-day basis. Examples might be on-time delivery, scrap rate, or individual process metrics.

Figure 5.1 is a visual depiction of the relationship between strategic- and operational-level metrics. Obviously the number and type of metrics can be expanded greatly for both strategic- and operational-level metrics. The figure is meant to represent a sampling of the many possibilities for selecting metrics.

Quantitative metrics are those that can be measured (variable data), counted, or are indicative of a particular state such as up/down, on/off, 1/0,

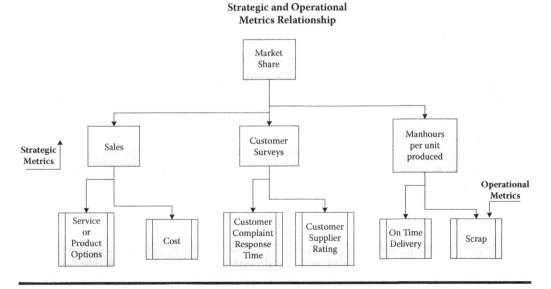

Figure 5.1

et cetera (attribute data). *Qualitative metrics* can come from diverse data such as observations, perceptions, descriptions, and comparisons. Although harder to assess, qualitative data can be just as important and sometimes more important than quantitative data. Often, tools such as Likert*-type scales are used to try to normalize and make more consistent readings of qualitative data from one measurement to the next. Both quantitative and qualitative metrics can be drawn from core or support processes. They can also be drawn from either external (customer) or internal sources.

To best determine which metrics to select for monitoring, you must first establish what information you are hoping to acquire, for what purpose, and what the sources of that information may be. Once you have established these facts, you need to set up your "metrics tree." A metrics tree is a lot like a family tree. Some metrics are closely related and others more distantly related. Some metrics drive others. Some metrics support others, whereas some are contrary. There may be a history to be aware of and you can often look back at some metrics and clearly discern where they originated.

If metrics are properly aligned then lower-level metrics regardless of whence they are drawn should support higher-level metrics. In other words, a significant shift in a lower-level metric should allow you to predict to some extent the impact on the higher tier metric, based on the relationship between the two. If a lower tier metric is not tied to a higher tier metric then that metric should be re-evaluated to determine continued usage. The reason for this action is that metrics should be tied to organizational goals and objectives, so if you have a metric that is not related to any other metrics, then by default, if you have chosen your other metrics well, the unaligned metric would not have any impact on organizational objectives and goals. And if that is the case, then why track it? Now let's look at some specifics of how different metrics may affect one another.

In his book *Performance Metrics: The Levers for Process Management*,[†] Duke Okes identifies three questions to ask when examining the relationship between metrics:

■ Are there significant gaps in what is being measured? That is, are there factors that are critical to performance for which there is not a metric?

---

[*] A Likert scale (usually 1 to 5 or 1 to 10) is a psychometric scale commonly involved in research that employs questionnaires and tries to quantify or assign a rating to different categories of responses, in order to promote consistency in evaluating one questionnaire to the next.
[†] Okes, D., 2013, p. 36.

- Is there misalignment between metrics? That is, does a specific metric drive one variable in the good direction and another in the bad direction?
- Is there conflict between metrics? This occurs when two (or more) metrics drive the same variable but in different directions when both are "improved."

See Figure 5.2 for a graphical depiction of each of these scenarios. The depiction is a modified version of an image taken from *Performance Metrics: The Levers for Process Management.** Driver, inputs, controls, or leading metrics can be used interchangeably, as can results, outputs, outcomes, or lagging metrics.

Once monitoring has been established and your metrics for success selected, then an organization must continually assess to verify that the metrics are providing the information for which they have been selected to provide. Another thing to be wary of is that often what is measured is what is easy, whereas what needs to be measured may be hard to calculate. Be mindful of this fact during the upfront selection of metrics. To paraphrase Kurt Stuke, operations audit director of Adecco, whom I recently heard at an ASQ Audit Division conference, "Quality is truth that can be measured." The road to delightful service, and the many benefits therein, goes through the land of "truth" in measurement.

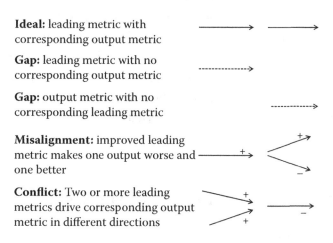

Ideal: leading metric with corresponding output metric

Gap: leading metric with no corresponding output metric

Gap: output metric with no corresponding leading metric

Misalignment: improved leading metric makes one output worse and one better

Conflict: Two or more leading metrics drive corresponding output metric in different directions

Figure 5.2 (Modified from Okes, D. 2013. *Performance Metrics: The Levers for Process Management*. Milwaukee, WI: ASQ Quality Press, p. 36.)

---

* Okes, D., 2013.

Case-In-Point: Importance of Monitoring

How we monitor is equally as important as what we are monitoring. In many standards and regulations, it is a requirement to report out of specification/tolerance conditions to area supervision, which makes sense. What doesn't make sense is that often the common sense corollary to that requirement—reporting out of control conditions—is often not addressed. For too many companies, it is often not a requirement or if it is, relevant staff haven't been adequately trained to recognize negative trends.

For one company an incident occurred where one of the components of a fabricating machine had started to wear over time. The corresponding dimension on the part being fabricated began to start measuring smaller and smaller, although still within specification. To add to the difficulty of the situation, there was a requirement that sampled parts be held for 48 hours before measurements are taken. This requirement created a two-day delay in process-monitoring data, making trend analysis an especially critical control to be in place.

As you might expect, eventually the dimension in question went out of specification and was found to have been out of specification for the previous 40 hours causing over $20,000 of scrap. Upon further review of the plotted data, it was seen that a downward shift had occurred three months previously and that the data had been trending downward since that time. Again, the data had been within specification and the C_{pk} had remained acceptable so no flags were raised. The C_{pk} is the process capability index which = [the upper specification limit (USL) − lower specification limit (LSL)]/six sigma (process spread). If the technicians taking the data had been trained to look for and sound the alert when negative trends appeared, this scrap loss could have been avoided.

Case-In-Point: Misalignment between Levels

For an example of misalignment of metrics, let's look at the case of a call center where appropriate metrics were selected but not properly utilized. At this call center two of the metrics that were measured were calls handled per hour and calls per hour that needed to be escalated to the next level. Employee A, having just recently come from training where she was told that it was important to retain a cordial relationship with the customer even though she was in collections, tried her best to answer questions and provide alternatives to customers within the realms of her responsibility. Employee A only escalated calls to the supervisor when requested or if she had exhausted all avenues in trying to reach an arrangement with the customer. Employee B, who had been with the company for a while escalated almost any call when there was a question, even when it was something she could easily look up. As one might expect, upon review of the quarterly metrics, it was found that Employee A was in the bottom third of calls handled per hour within an average month, but had the fewest escalated calls (and fewest complaints). Conversely, Employee B had the second highest ranking of calls handled per hour. Employee B was awarded one of the small quarterly performance bonuses that were handed out and Employee A was given a verbal warning.

Employee A was told that "it was nice" that she had the fewest escalated calls but that they were all there to process calls and collect funds. Calls should be processed as quickly as possible even if that meant allowing the customer to take a payment option that might cost them more money and stress, rather than fully explaining all of the options. She was also told in the same discussion that "if she wanted to help people, she should go to customer service." When Employee A inquired how to transfer to the other department, she was told that she was hired for collections. If she now wanted to work in customer service, she would have to quit and reapply for the other department. Employee A did quit shortly after that discussion but she didn't bother reapplying. Somewhere along the line the company forgot to make sure that the meaning of their slogan of "We never stop working for you" flowed down to the collections department. This misalignment of priorities, between strategic and operational levels, in this case evidenced by improperly weighting the metrics collected, cost the company a good employee and will no doubt lose them many customers and possibly market share, in the long run.

Chapter 6

Root Cause Analysis

Introduction

Now after you have selected the appropriate metrics, are monitoring and reporting them to upper management, what do you do when they are telling a story that you are not happy with? At the heart of every Lean, Six Sigma, and continuous improvement project is one thing: root cause analysis. After all, you can't do design of experiments (DOE) until you know what you are trying to prove. Root cause analysis is often thought to be only used for fixing problems. It can also be used to determine why the current state doesn't match the desired future state. Through implementation of the Kano philosophy, two ways to achieve delightful delivery are seen: constant striving plus structured and incremental improvements based on analysis, identification of an issue or opportunity, and evolution of practices. In order to improve, the impediments to improvement must be removed. Root cause analysis is the process of determining the root cause of a particular event. This event can have occurred in the past or be expected in the future; the event can be either positive or negative. *Root cause* is the first most principal cause of an event. Once potential root cause has been determined, it must be confirmed before either corrective or improvement action can take place. *Corrective action* is that which when taken, eliminates or significantly reduces the impact of the root cause of a problem, thereby preventing or mitigating the effects of recurrence. *Improvement action* is the implementation of a planned improvement initiative in order to reach a desired goal. It is this type of ongoing improvement action that will drive an organization toward delightful service delivery. Whenever possible, corrective and improvement actions should be tied to quantifiable metrics so that you have an easy way to see if the action taken has been successful. Quite simply, if the metric moves the desired amount in the right direction, then you have been successful.

Some of the skills needed for root cause analysis are observation, listening, analysis, team work, and communication for reporting the results of the investigation. These are all skills that can be cultivated. They are important but not difficult to learn. Root cause analysis can be simple or complex. Training in root cause analysis should be tailored to the level of expertise of the trainee. All levels of the delightful organization will have had some form of root cause analysis training.

The 5-Whys is a simple but powerful technique. The way it works is that once an issue is identified, the investigator asks "Why" until there is no answer to the question. When that point is reached, the investigator has arrived at the root cause of the issue. Even though the process is called the 5-Whys, it can be 7-Whys, 3-Whys, or any number of whys that it takes to determine root cause.

Let's take the example of an overhead projector that is not working because it has been unplugged, to see how this process works.

Why is the projector not working?
Because it is not plugged in.
Why is it not plugged in?
Because the cable has been pulled out of the outlet.
Why was the cable pulled out of the outlet?
Because something got hooked onto the cable and pulled it out.
Why did something get hooked onto the cable?
Because the cable lies on the floor and gets in the way.
Why does the cable lie on the floor and get in the way?
Because it is too long.
Why is the cable too long?
Hmmmm? ... I don't know ...

So in this example, the root cause of the problem is that the projector cable is too long. This is the problem that must be addressed in order to prevent a recurrence of the problem.

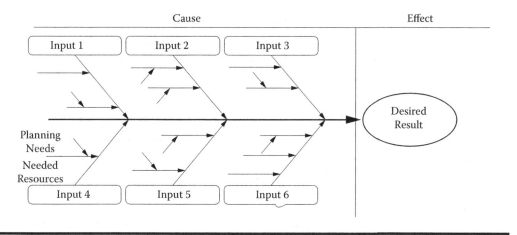

Figure 6.1

When rolling out an organizationwide improvement program this is the method that all employees should be trained on because of its ease of understanding. Then more skilled employees may be trained in more complex methods of root cause analysis. Another widely used method for determining root cause is the Ishikawa diagram, developed by Kano's mentor Kaoru Ishikawa. As discussed in Chapter 3, the Ishikawa diagram is also called the fishbone (after its shape) or cause-and-effect (after its function) diagram. The cause and effect diagram, as seen in Figure 6.1, can be populated after brainstorming with possible drivers of the desired effect. Then experimentation and data analysis will take place to determine if the desired effect can be achieved.

The process for determining root cause is as follows:

- Identify opportunity/gap.
- Determine possible reasons that the gap exists.
- Classify the opportunity or gap.
- Determine the magnitude/spread of the gap.
- Identify most likely reason.
- Confirm.

It is important to recognize the distinction between fact and opinion when going through the root cause analysis process. Let us use the following simple example where statement 1 is 100% true. Statement 2 is based on statement 1, and is told by Mr. Smith's assistant to another party who is also going to be in the meeting.

Statement 1: Chris Jones has a morning meeting with Mr. Smith and calls to tell him that he is caught in traffic and is going to be late.

Statement 2: Mr. Jones called Mr. Smith from his cell phone because he was caught in traffic and thinks he is going to be late.

Fact	Opinion
Chris Jones is caught in traffic	Chris Jones could be either male or female; the speaker is assuming male.
	Most likely called from cell phone but might have had to pull over and call from a land line.
	Chris might think he or she will be on time, but is calling just in case. We can't know what he or she is thinking.

In nine out of ten cases, such subtle distinctions might not matter, but when investigating root cause, the investigator should be aware of such differences for that tenth time, when the distinction might move the investigation either away from or toward the true root cause being sought.

Determining root cause means assessing what is keeping you from moving forward, or what the difference is between your current state and your desired future state. This is a skill that must be entrenched within the organization embracing continuous improvement in pursuit of delivering delightful service.

A helpful tip when thinking about the process of determining root cause is to think of 5 Ws plus an H. So what do I mean by that?

What – This is your problem statement. What is the problem? What is it that we would like to see improved?

Where – Where (physical location) is the problem taking place? Where would you like the improvement to take place?

When – Time-based information, asking when does the issue occur: season, shift, day of the week, and so forth. When is the best time to implement the improvement? When are you most likely to achieve success?

Weight – How bad or expansive is the problem? How big an impact can the improvement have on the organization's operations?

How – How did the problem escape our controls? How can we standardize the improvement, in order to sustain the benefits?

Why – Why did the problem occur (root cause)? Why does this opportunity exist?

Ask the first four Ws and they will drive you toward the H and final W, identifying root cause and preparing you to begin developing corrective or improvement action plans.

Another way to think about root cause is to look at seemingly identical situations such as production lines or telephone operators (in that they have been trained the same and have the exact same responsibilities) in order to distinguish similarities and differences. Then you ask yourself why there is a problem or opportunity with one and not the other. In other words what is one, and what is not the other? It is this distinction that will point to the root cause of the issue or opportunity.

One more useful ability of root cause analysis is in determining why something worked. Often an organization may want to translate the benefits from an improvement into other areas of their operations. First there has to be a recognition that there is the potential for translation and then determine if a transferring improvement action can take place.

Case-In-Point: Is/Is Not

Let's look at the case of a call center that has seen an increase in two successive quarters in the number of calls that need to be escalated to the first level of management as shown in Figure 6.2.

During the preceding year, the company expanded its operations and consequently had twice as many new hires as would be typical in a given year. Most of these new hires were placed on the second and third shifts as shift placement is done by seniority and the preferred shift is the first shift. New training and procedures were also put in place the previous year, with

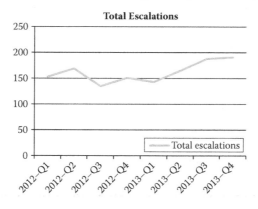

Figure 6.2

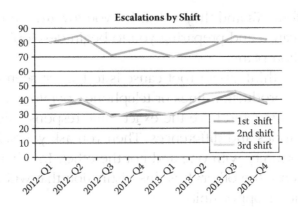

Figure 6.3

the intent of reducing the number of escalations, and so the office manager is surprised to see data reflecting an opposite trend. The site goal is to have no more than 10% of calls escalated and as of the last two quarters, that goal is no longer being met. The office manager suspects that the larger than usual number of new employees might be the reason for the escalation in calls, but wants to confirm his suspicion and so begins conducting a root cause investigation with his team of three shift supervisors. The first thing that the office manager does is to sort the data by shift to see if that will tell him anything. Please see Figure 6.3.

At first glance, it seems as if there may be a problem in the first shift, but more analysis remains to be done. The second thing to be done is to "normalize" the data. By looking at the percentage of calls escalated, this eliminates higher call volume as the primary cause of the increase in escalations (Figure 6.4).

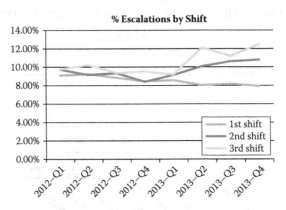

Figure 6.4

Calls Escalation Cause-and-Effect Diagram

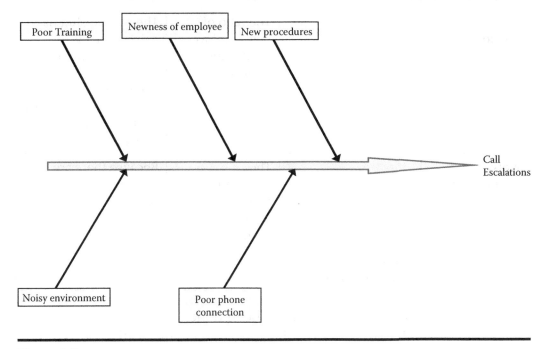

Figure 6.5

Further review of the data yielded two pieces of information. Although the number of escalations increased for all three shifts, the actual percentage of calls escalated did not. Per the data reviewed, the first shift is trending downward in the percentage of calls escalated whereas the second and third shifts are trending upward. Next the team brainstorms possible reasons for the escalation and plugs that information into a cause and effect diagram as seen in Figure 6.5.

Now we are ready to do an IS/IS-NOT analysis to see what is happening differently on the first shift from the second and third.

Potential Cause	IS	IS NOT	Comments
Training adequacy	The same training IS given to all shifts by the same trainer.		
New procedures	The new procedure IS implemented across all shifts at the same time.		

Potential Cause	IS	IS NOT	Comments
New employees		The ratio of new employees IS NOT the same across all shifts.	The second shift has twice as many employees with less than one year experience than the first shift and the third shift has three times as many.
Noisy environment		The environment IS NOT exactly the same for each shift.	Because there are more people around on the first shift, that shift tends to be noisier.
Poor phone connections	The same phone system IS used by all.		

Now that the IS/IS-NOT analysis has been done, we look at the differences between shifts and note that there are two relative to the identified potential causes. The first difference is that there are noticeably more new employees on shifts two and three as opposed to shift one, which has been improving over time. The second difference is that shift one tends to be significantly noisier than shifts two and three, which might have an adverse effect on their performance. Looking once again at the data, we see that shift one is outperforming shifts two and three despite the noisier environment, so we discount environmental noise as a potential root cause along with training, the new procedures, or poor phone connection.

It is determined that further investigation should be conducted to determine why the greater number of new employees on shifts two and three might be causing an increase in escalations. The 5-Whys would be one good tool to use. Then once a theory has been arrived at, the next step would be for the office manager to prove or disprove that theory. This is how IS/IS-NOT points the way and can be used alone or in conjunction with other tools to investigate root cause.

Chapter 7

Project Management

Introduction

For members of a truly delightful organization, project management skills are essential. These skills allow project teams to set measureable goals and success criteria as well as (and just as important) criteria for when to abort a project. They also help keep the team focused and manage time and resources, along with properly documenting the project. A project can be as simple as cleaning one's closet or as complex as installing a multimillion-dollar piece of equipment. There are five phases to a project—initiation, planning, execution, monitoring and controlling, and closure.

Initiation: This refers to selecting the project goal. It is in this phase that the project charter is developed. It is also critical that in the development of the charter, the project scope is clearly defined, benefits and risks identified, and management buy-in confirmed.

Planning: Planning the project involves estimating needed resources, identifying team members, developing a project timeline, metrics, reporting protocol, and implementation strategy.

Execution: This involves performing the tasks, documenting results, reporting progress, and making modifications to the plan as necessary

Monitoring and Controlling: This actually occurs during all phases of a project. It includes monitoring project resources, timeline, and deliverables, and providing feedback on overall project status.

Closure: This includes the assessment of project results against project goals, generation of the final report, and all other activities required for closure of the project.

Development of the project charter is the most important part of the initiation phase. The purpose of the project charter is to:

- Clearly define the project scope, or project boundaries. Many projects fail because the scope is too broad or the boundaries are undefined. Often a SIPOC diagram (as previously discussed in Chapter 4) is used to illustrate clearly the project boundaries through identification of start and end points as well as inputs, suppliers, outputs, and customers for a given process.
- Establish measurable objectives, or in other words, identify a quantifiable gap. To use the case-in-point from the previous chapter, the call center's goal was fewer than 10% of calls escalated. The average percentage of calls escalated for the previous two quarters was 10.7%. The gap to close is 0.8% calls escalated across all shifts.
- Identify project benefits. It is important not only to identify measurable objectives but also to state plainly why meeting these objectives is important and what benefit the organization will attain by successful project implementation.
- Identify project risks. Project risks must be identified and assessed, and possible mitigation explored.
- Identify needed resources including personnel, funding, equipment, time, and so forth.
- Provide justification for the project in order to obtain management buy-in. Project benefits as well as the likelihood of success must be weighed against any residual project risk.

It is the project charter that is referred to in order to keep the project from going off track. Documenting project activities in order to capture lessons learned and to provide a record of activities that can be referred to in the future is very important. Project documentation can include any or all of the following and more:

- Project charter
- Meeting minutes
- Interim and final reports

■ Correspondence
■ Data that were captured and analyzed
■ Executive summary of project results

Project documentation should be stored in a manner that is easily accessible for a time previously determined by organizational procedure.

Project reports should be distributed as specified in the reporting protocol when key milestones arrive, upon times of significant change, and upon completion of the project. As a minimum reports should contain:

■ Executive summary providing overall project status
■ Identification of project team lead and team members
■ Specific status with regard to project deliverables and timelines
■ Any major or unexpected changes in the project
■ Whether goals and objectives are being met
■ Charts, graphs, or other illustrations
■ Relevant project data

A project can be implemented by a team of one or by a cross-functional team of five to nine people. The three most important factors to consider when selecting team members are the skills needed to complete the project successfully, the time and commitment needed to participate in the project, and finally, how well the team members will work with one another. One other important consideration is to have, when possible, a member of the team who is not vested in the project success, thus providing a more impartial opinion.

Also important to successful project completion is good communication. Three rules for successful communication are:

■ Ensure that the message has been received.
■ Ensure that the message has been understood.
■ Use the appropriate communication medium.

The first two statements are fairly obvious but, as discussed before, what do we mean by using the appropriate communication medium? Well, if you e-mail someone and they always respond by calling you back, then the phone is more than likely their preferred method of communication. Important communication then, should be done by phone, with a follow-up in writing. It is important that this follow-up summarize what was discussed and identify any decisions made as well as any applicable action items and deadlines.

Implementing a project may entail certain risks. Risk is defined by ISO 31000:2009 as the "effect of uncertainty on objectives." Three broad categories of risk to be aware of when implementing a project are below:

- Using resources and not being successful
- Implementation of the project bringing risk into the environment in which the project is unfolding
- Implementation of the project resulting in worse results

Once the risks have been identified they need to be assessed. Two simple risk assessment tools are the *risk priority number* (RPN) and the *risk matrix*. Quite simply the risk priority number assigns a value to each of three risk components related to a specific risk, from 1 to 10 with 10 being worst case. Those three categories of risk are:

- How likely is it for this adverse event to occur?
- If it happens, how bad will things get?
- If it happens, how likely are we to detect it, and how soon?

Then the numbers are multiplied out and the answer is the risk priority number.

For example, say there is a problem that if it occurs, the ramifications would be very severe. We would rate it a 9 on the severity scale. If it is somewhat likely to occur, it would rate a 5 out of 10 on the likelihood scale. However, if we can easily detect it as soon as it occurs we would give this a 2 on the detection scale (if it were hard to detect, we might give it an 8). So the risk assessment for this scenario would be calculated as:

$$RPN = 9 \times 5 \times 2 = 90$$

Another simple tool is the risk matrix which is a color-coded matrix that has impact ratings across the top and frequency ratings along the side as shown in Figure 7.1. Then the team or individual assigns a risk code (e.g., green, yellow, or red) according to the definitions assigned to the matrix, and pursues actions as dictated by organizational procedure.

Risk mitigation is the attempt to eliminate a particular risk or, if it can't be eliminated, to minimize the impact. Once risk mitigation has occurred, then the risk is assessed again, and any remaining risk is considered the residual

Frequency	Impact					Recommended Action
	Negligible	Minor	Important	Critical	☐	Initiate a product recall
Continually	☐ M	☐ H	☐ H	☐ H	☐	Open CAPA Investigation (Approval required for below actions)
Frequently	☐ L	☐ M	☐ H	☐ H		
Occasionally	☐ L	☐ M	☐ M	☐ H	☐	Write NCMR and Investigate
Rarely	☐ L	☐ L	☐ M	☐ M	☐	Correct and Document
	Low Risk (L)	Medium Risk (M)	High Risk (H)		☐	Track and Trend

Figure 7.1

risk of the project. At that point the benefits of the project are weighed against any residual risk and a determination is made whether to move forward with the project. It is up to the team to determine what is considered an acceptable rating, either on their own or through interpreting an existing corporate standard. For complex projects, a risk assessment should be done at the start of the project as well as at strategic points on the project timeline.

Some important skills that are needed when managing a project are organizational, communication, attention to detail, time management, and even negotiation, as well as the ability to get along well with others. Some character traits that are important for the successful project manager are tenacity, determination, open-mindedness, inquisitiveness, fairness, and honesty. Now, let's take a quick look at some of the tools that would be used to help manage a project in its various phases.

Initiation/Planning

- Cause-and-effect diagram: Choosing project
- Project planning template: Planning
- Gantt charts: Planning
- SIPOC diagram: Planning
- Decision tree: Project justification
- Expected profit formula: Project justification
- Risk priority number/risk matrix: Risk management

Execution

- Data collection tools: Feedback
- Critical path method: Task prioritization

- Work breakdown structure: Structured list showing how tasks are related and will be accomplished
- Risk priority number/risk matrix: Risk management

Control and Monitoring

- Control charts: Data collection
- Measurement system evaluation analysis: Verifying accuracy of data
- Risk priority number/risk matrix: Risk management

Closure

- Project report: Summary and conclusion

Typical metrics attached to projects are related to actual versus expected expenditures such as:

- Percentage of work completed on time
- Percentage of work completed on budget
- Percentage of resources used versus percentage of project completed
- Actual results (to date) versus expected

Also to be tracked is some metric tied to the end goal of the project. In other words, are you starting to see the "dial move" in the direction that you wish? It is this type of structured approach that will lead to ongoing success in organizational improvement initiatives. To conclude, you can't drive continuous improvement in pursuit of delight without a solid project management skill set.

Chapter 8

Putting All the Pieces Together

Delight in Manufacturing

So what would an organization (or individual) have to do to prepare itself to deliver delightful service? First, there must be a clear vision validated by research and study. The organization must know its environment: who are the leaders, what are their strengths, and why? The organization must reflect internally in order to know its processes thoroughly and understand where value is created within them. The organization would further have to understand its unique value proposition and how to build upon it. In other words how do we leverage our strengths? The determination of a unique value proposition is in respect to internal as well as external customers. There must be the ability for self-reflection and honest self-assessment.

There must also be a certain mindset, a built-in desire for continual improvement and a yearning to deliver delightful service and to receive the many benefits from doing so. From there, the rest is implementation.

How do all of these tools work together in providing customer delight? Let's take a look at the case of a fictional manufacturer of high-end quality wooden furniture that wants to expand its business into a new market. It decides to leverage existing skills and supplier relationships to begin manufacturing wooden walking canes. With the population of the United States aging, there is expected to be 20% annual growth in that market for the foreseeable future. Currently the largest player in the market owns

Strategic Mission Deployment – X-Matrix

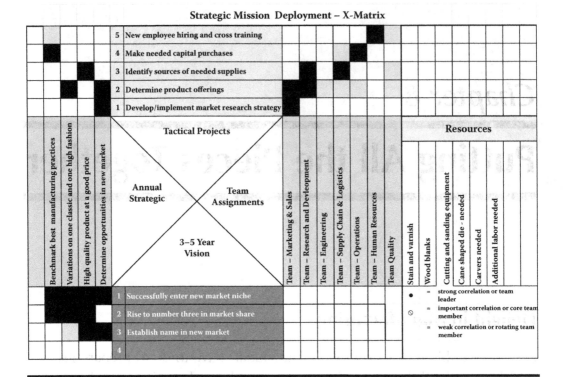

Figure 8.1

40% of market share, the second 25%, and the remaining 35% of the market is split among 12 other companies. After arriving at a strategic vision, surveys and market research are conducted in order to assess the potential to provide unique value to the customer as well as the viability of pursuing this path. Based on feedback from the surveys, it is decided that the initial offering would consist of two canes: one with a standard curved handle and one with a carved handle, both with a light and dark stain, and each one having a rubber footer.

Then a Hoshin Kanri X-matrix is developed in order to align everyday operating parameters to annual strategic goals to the overall strategic plan. The first level shown in Figure 8.1 outlines the three- to five-year plan showing strategic vision/mission, strategic goals, tactical objectives, and team responsibilities in addition to available and needed resources.

Note: The headings of the Hoshin Kanri X-matrix can vary according to the organization or preferred method of implementation. It is most important that the matrices reflect and flow down information from the long-term strategic to annual strategic plans, to ongoing operations.

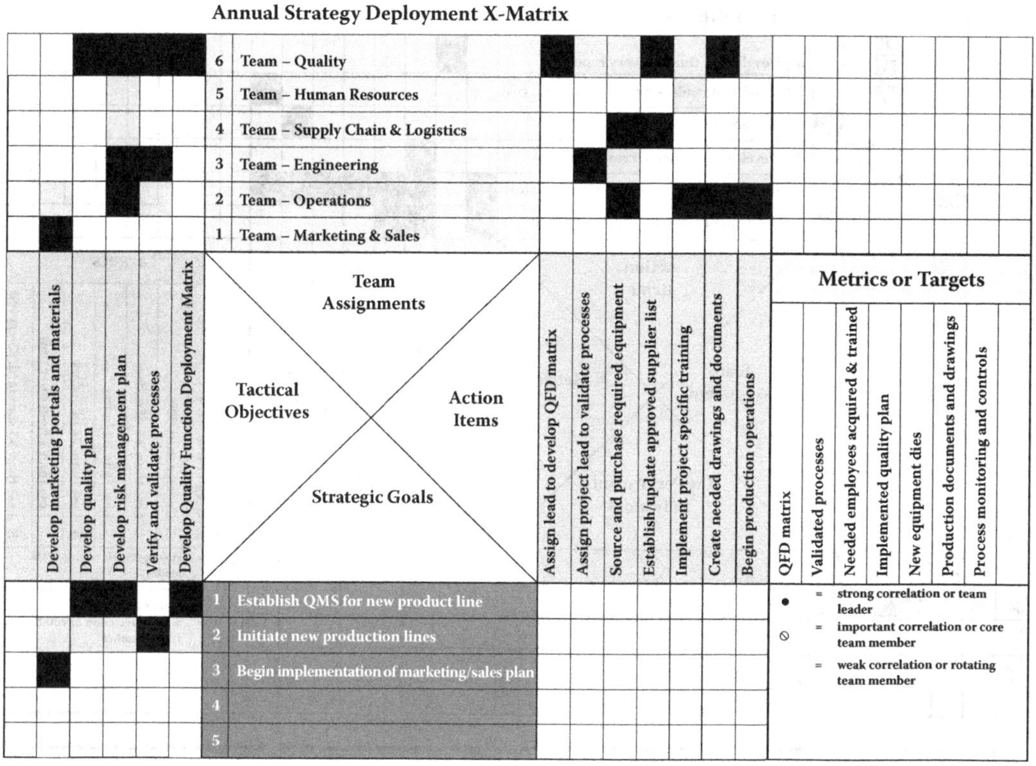

Annual Strategy Deployment X-Matrix

6 Team – Quality
5 Team – Human Resources
4 Team – Supply Chain & Logistics
3 Team – Engineering
2 Team – Operations
1 Team – Marketing & Sales

Develop marketing portals and materials
Develop quality plan
Develop risk management plan
Verify and validate processes
Develop Quality Function Deployment Matrix

Team Assignments

Tactical Objectives

Action Items

Strategic Goals

Assign lead to develop QFD matrix
Assign project lead to validate processes
Source and purchase required equipment
Establish/update approved supplier list
Implement project specific training
Create needed drawings and documents
Begin production operations

Metrics or Targets

QFD matrix
Validated processes
Needed employees acquired & trained
Implemented quality plan
New equipment dies
Production documents and drawings
Process monitoring and controls

1 Establish QMS for new product line
2 Initiate new production lines
3 Begin implementation of marketing/sales plan
4
5

● = strong correlation or team leader
◒ = important correlation or core team member
  = weak correlation or rotating team member

Figure 8.2

The second level outlines the activities needed to achieve the strategic goals and tactical objectives for the first year of the strategic plan. Please see Figures 8.2 and 8.3. These activities also include development of a quality function deployment (QFD) matrix in order to align customer explicit and implicit needs with manufacturing goals and production requirements. An action plan is developed for each action item identified in the matrix.

Figure 8.4 contains a sample action plan relating to development of the quality function deployment matrix identified as one of the first-year tactical objectives.

The quality function deployment matrix as seen in Figure 8.5 is then designed to correlate requirements with both explicit and implicit customer needs. Needs are ranked and prioritized so that resources can be appropriately allocated. This upfront planning allows for the opportunity to provide not just a quality product but to continue to try to delight the customer. As feedback is received from the field, the quality function deployment matrix is updated to include market perceptions as well as sales and other data.

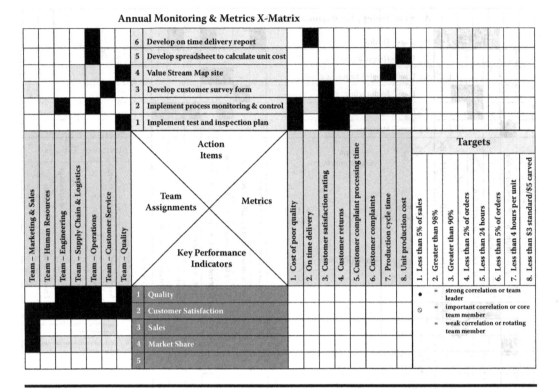

Figure 8.3

Before beginning production, suppliers are identified, processes are mapped, resources and training needs are identified, critical control points established, and a plan for monitoring and metrics developed. Process maps and flow charts are developed during the planning phase and value stream maps are developed during implementation.

The steps of the manufacturing process are seen below as well as in the flow chart (Figure 8.6) and process map (Figure 8.7):

1. Procure raw materials based on approved designs.
2. Inspect raw materials at incoming.
3. Cut wood.
4. Inspect.
5. Sand.
6. Inspect.
7. Carve handle.*
8. Inspect.*

---

* These steps are only for the canes with the carved handle.

Action Plan																	
Improvement Priority: Red				**Management Owner:** Quality												**Date Created:** 12 Dec 12	
Team: Quality Engineer, Operations Manager, Engineering Manager, Customer Service																**Next Review:** 31 Mar 13	
Background: QFD matrix needs to be developed for new product line																	
Relation to Annual Objective: Develop QFD Matrix				**Timeline**												**Status** Red, Yellow, Green	
				= Original Plan x = Complete													
				Planned Complete Date	**2014**												
Action Step/Kaizen Events	**Owner**	**Deliverable**			Jan	Feb	Mar	Apr	May	Jun	Jul	Aug	Sep	Oct	Nov	Dec	
Make a list of customer needs	Q/O/E/C	List of needs		01/31/13	x												
Match functional requirements to individual needs	E	Functional requirements		02/28/13		x											
Create specifications to to functional requirements	E/Q/O	Specifications and drawings		04/30/13				x									
Develop matrix template	Q	QFD Template		02/28/13		x											
Develop matrix feedback model	Q	Matrix input		04/30/13				x									
Inout data into matrix	Q	Completed draft matrix		05/31/13					x								
Review matrix	Q/O/E/C	Reviewed matrix		06/15/13						x							
Present QFD matrix to management team	Q/O/E/C	Final QFD Matrix		06/30/13						x							
Provide external feedback to quality	C	Ongoing		Ongoing													
Modify QFD model based on internal/external feedback	Q	Ongoing		Ongoing													
Q = Quality O = Operations E = Engineering																	
C = Customer Service																	

Figure 8.4

9. Stain.

10. Varnish/seal.

11. Apply footer.

12. Final inspect.

Raw materials and goods needed for the process are stated below and also reflected in the process map.

■ Wood

■ Stain

■ Varnish

■ Rubber footers

■ Different grades sandpaper

The established process must be monitored via the reporting of previously selected metrics reported to various levels of management as well as through regularly scheduled audits, plus executive-level Gemba walks. External

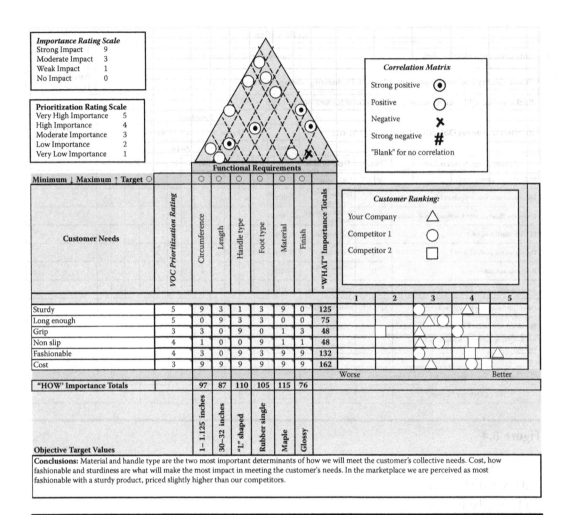

Figure 8.5

feedback, such as from customers, market surveys, Internet portals, and so forth, must also be continually sought and fed back into the organization.

Now that we have established an effective operation, we must look to becoming as efficient as possible. To that end, we develop a value stream map of the current state. This will allow us to identify production bottlenecks as well as value- and non-value-added activities. The lead time and value-added times reflect averages just for the sake of example. Looking at the simple value stream map in Figure 8.8, one can see opportunities to reduce or eliminate waste through lead (waiting) time, transit (shipping) time, and error (defect %). Not as obvious are opportunities to become more efficient and thus utilize less time and fewer resources during the value-added as well as receiving, inspection, and shipping activities.

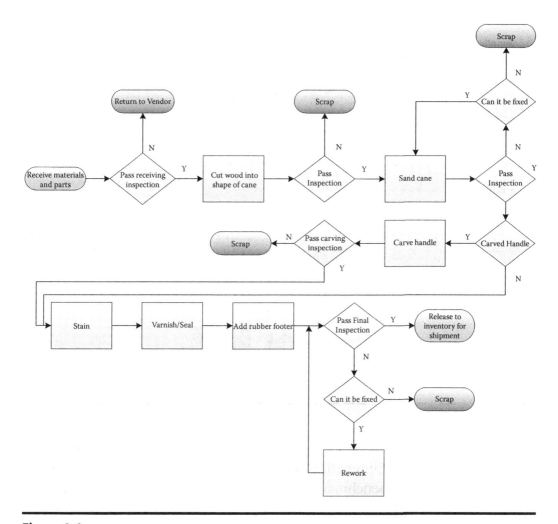

Figure 8.6

Notice that if you multiply the yield from each step from Figure 8.8,

$$\text{Total Yield} = 0.90 \times 0.95 \times 0.90 \times 0.95 \times 0.90 \times 0.90 = 0.592$$

The output is less than 60% of your theoretical yield if everything is perfect and no items are rejected in any of the steps. Similarly, when you compare the time within the plant spent on value-added activities versus the time spent on non-value-added activities in Figure 8.8, you discover that:

% Time Value-Add = number value – add hrs/total number hrs = 24/67 = 35.8%

Only one third of the time spent within the plant to process a shipment of canes is spent on value-added activities!

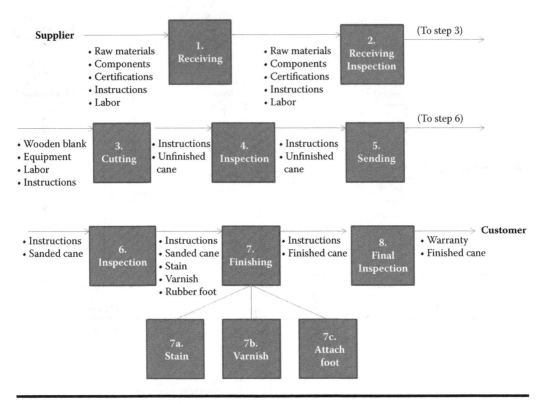

Figure 8.7

Improving yield can lead to not only increased profits, but also reduced cost for the end user. Eliminating time spent on non-value-added activities frees up resources to benchmark best practices both internally and externally as well as to pursue other continuous improvement opportunities. So even in an operation that is perceived as efficient, there can be huge opportunities to eliminate waste and provide increased value to the customer. How delightful would an unexpected reduction in both lead time and product cost be to your customer? Now you take another look at the map to ensure that the "why" and the "how" of each waste makes sense and also to see if you can think of any other potential wastes in the system, as it is described.

Several things can be learned by looking at this exercise in its entirety. First of all, the addressing of the customer's implicit and explicit needs is seen from the initial highest level strategic planning down to the day-to-day manufacturing activities. The process monitoring activities and developed metrics are both appropriate, not just to the day-to-day operations, but also for the long-term strategic goals as well. The organizational mission and long-term strategic vision have successfully flowed down via annual

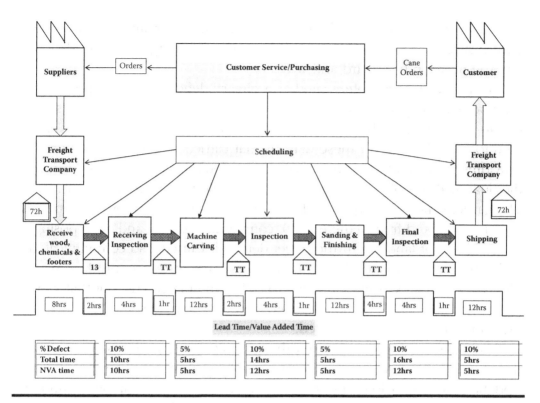

Figure 8.8

strategic goals and tactical objectives from the boardroom to the production floor. Both internal and external data are reviewed to provide feedback so that negative trends can be addressed, positive trends can be encouraged, and so that plans can be revised as necessary.

Delight in Service Provision

Latrice recognizes that there are homeless and struggling people in her community and wants to become involved in the efforts to feed the needy. She decides to form her own nonprofit organization to work within this arena. From previous research Latrice knows that this is easier said than done; it has been reported that although there remains a pressing unmet need for food donations, almost 40% of the fresh food that is collected for the homeless and disenfranchised spoils before it is delivered.

Part of the decision-making process in starting her nonprofit organization is whether to collect and deliver uncooked food, cooked food, or both. Also she must decide on whether to focus on certain food categories such as vegetables or canned goods, or will her organization do it all. Latrice has to decide what skill sets she will need to bring into the organization in order to ensure success as well as how she will structure the organization. Also, who will she need on her board and why?

Fresh vegetables are often cited as a dietary deficiency in the poor, typically because they are relatively more expensive than processed foods, and that in some parts of the city (where there are no supermarkets) they are hard to come by. It is for these reasons that Latrice decided to focus on providing fresh produce to the needy. This would be accomplished through the dual methods of community gardens and donations from area farms. The gardens met the purposes of ease of access, low cost (for the nonprofit to implement and free to the needy), community buy-in, and repurposing of unused plots of land. To further meet demand, Latrice also communicated with small area farms to pick up their after-market produce, produce with blemishes or maybe a little too ripe to sell to the big chains, but that were perfectly edible. She assembled a core team of volunteers and even got an older but functioning car donated.

Latrice also had to determine if the organizational model of operation that she had decided upon would be feasible to implement. Storage, transportation, and other logistics all had to be addressed in addition to funding for the fledgling organization. Once these decisions had been made, the Growing-in-Grace (GIG) nonprofit was established. Read on to follow the steps along this journey.

The first step was to figure out what she needed from her executive board beyond just oversight. With that in mind Latrice used an inverse Ishikawa diagram to illustrate visually what skills and resources she needed on her board before setting out to invite community leaders to support her nonprofit as executive board members (Figure 8.9). She determined that five would be a good number because it would allow her to have as broad a cross-section of guidance as possible without the number being so unwieldy that it would be hard to reach a consensus when decisions had to be made. Also, Latrice thought that if board members became unavailable on occasion, a quorum of three should be fairly easy to arrange for a vote on an important issue that couldn't wait for availability of the full executive board.

Once she determined what skills and resources were needed in her leadership team using the inverse Ishikawa diagram, she went about seeking individuals with those skill sets or connections. Latrice recognized that even though this was a nonprofit organization, those she was helping were still her customers and should be considered as such during the planning process. She also recognized that a quality standard had to be set even for

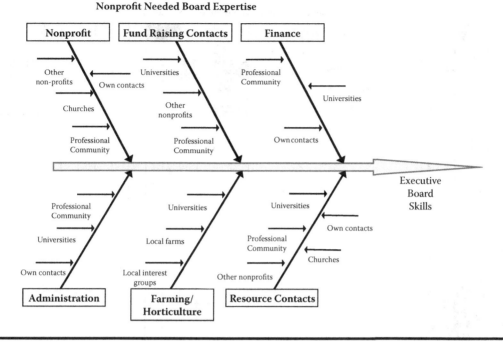

Figure 8.9

items that were being donated. So once the executive board was selected, Latrice sat down with them to develop a Hoshin Kanri X-matrix to ensure that customers' needs and the quality of the service provided were taken into account in planning for all levels of the organization. The matrix having the highest level of strategic planning is shown in Figure 8.10.

The black, white, and shaded boxes indicate primary, shared, and nonresponsibility as well as where certain groups will have to work together. Latrice, as director, will establish and review performance metrics regularly to ensure success, and provide status reports to the executive board.

As part of the Hoshin Kanri planning process, Latrice and the board then met to brainstorm the needed organizational functions to ensure the success of their mission, and to determine the best organizational structure to implement this organizational vision. The cluster diagram in Figure 8.11 is the result of this first brainstorming session.

The action plan shown in Figure 8.12 was used to document a structured methodology with milestones and timelines for going through this process.

After all possibilities had been exhausted, Latrice and her board decided to combine functions where it made sense and develop the organizational chart from there. The consolidated cluster diagram and resulting organizational chart are seen in Figures 8.13 and 8.14.

Strategic Mission Deployment – X-Matrix

				5	Identify potential partners																			
				4	Recruit volunteers and volunteer leaders																			
				3	Recruit board members																			
				2	Develop organizational structure																			
				1	Develop stategic plan																			

Benchmark existing best practices	Identify operational gaps	Identify needed subject matter experts	Identify strategy and craft mission statement	Tactical Projects			Executive Board	Director	Team – Operations	Team – Supply Chain & Logistics	Team – Volunteer Coordination	Team – Strategic Partnerships	Volunteers	Employees	Board Members	Land	Nonprofit Foundations	Strategic partners

Tactical Projects

Annual Strategic — Team Assignments

3–5 Year Vision

Resources

				1	Successfully enter new market niche								
				2	Develop model that can be transferred								
				3	Establish strategic partnerships								
				4									

● = strong correlation or team leader
◌ = important correlation or core team member
= weak correlation or rotating team member

Figure 8.10

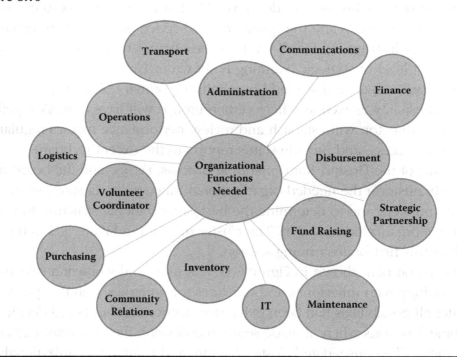

Figure 8.11

Improvement Priority: Red			Management Owner: Director														Date Created: 12 Dec 12
Team: Director, Operations Manager, Executive Board (review and approval)																	Next Review: 31 Mar 13
Background: Develop Operating Procedures and Organizational Chart																	

Relation to Annual Objective: Implementation of Strategic Plan			Timeline														Status Red, Yellow, Green
				= Original Plan		x	= Complete										
			Planned Complete Date	2014													
Action Step/Kaizen Events	Owner	Deliverable		Jan	Feb	Mar	Apr	May	Jun	Jul	Aug	Sep	Oct	Nov	Dec		
Make a list of customer needs	E/D	List of needs	01/31/13	x													
Match functional requirements to individual needs	E/D	Functional requirements	02/28/13		x												
Make a list of needed organization functions	E/D	Functional outline	04/30/13				x										
Develop organizational chart	E/D	Organizational chart	02/28/13		x												
Develop job descriptions	D	Formal job descriptions	04/30/13				x										
Recruit staff	E/D	Full staffed organization	05/31/13					x									
Develop operational procedures	E/D/O	Operational Procedures	06/15/13						x								
Develop reporting protocol	E/D	Embedded in policies & procedures	06/30/13						x								
Train staff	D/O	Ongoing	Ongoing														
Develop metrics for success	E/D	Ongoing	Ongoing														
D = Director O = Operations Manager																	
E = Executive Board																	

Figure 8.12

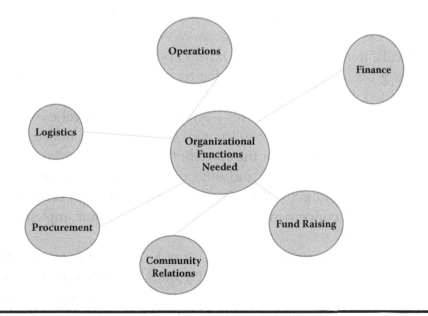

Figure 8.13

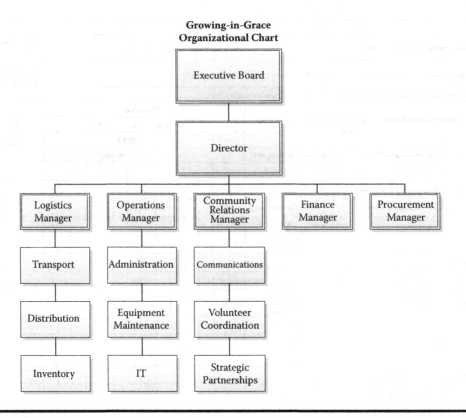

Figure 8.14

Next, flowing down from the strategic level, a Hoshin Kanri X-matrix must be developed to establish and document annual goals and objectives that support the long-term strategic goals of the organization (Figure 8.15).

And finally, performance metrics have to be established and tracked to ensure that goals are being met and that the results of meeting those goals are as expected. Please see Figure 8.16.

By flowing down its strategic mission through each level of the organization, GIG ensures that the primary goal of delivering a needed service to the community is always in the forefront of everyone's consciousness. By including measurable goals and objectives, Latrice makes it harder for the organization to get "off track" and provides the opportunity to recognize areas of concern should they arise and prompt implementation of needed changes.

Finally, as a reminder that their goal is not just to deliver good service, but delightful service, a Kano diagram (Figure 8.17) is placed on the wall, with organizational placement on the diagram, dependent upon regular self and customer assessments.

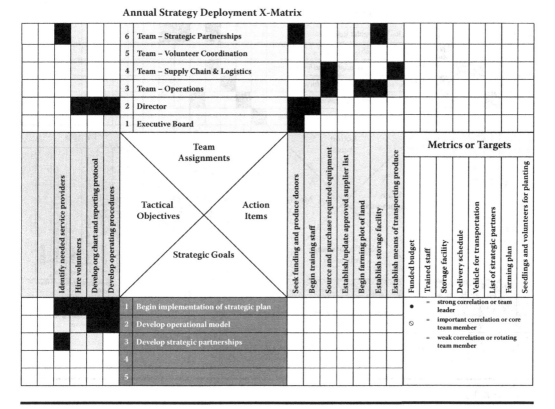

Figure 8.15

Depending on where they are on the model will determine the questions that need to be asked for the organization to continue to improve in the service that they provide. For example, if the overall assessment landed GIG in the bottom two quadrants, then the questions to be asked are what are we doing wrong, what can we do to correct it, and how can we do better.

When in the top left quadrant, the organization must ask, how can we benchmark best practices and what are the continuous improvement initiatives that we can put in place. When driving the organization toward the assessment of delivering delightful service as shown in the top right quadrant, the questions that must be asked include what are the unspoken needs of the customer and what is something new, that hasn't been tried, that could lead to breakthrough performance?

Latrice implemented this final operational step, because she understood that at the heart of delivering delightful service is constantly asking, "Am I delivering excellence, continuous improvement, and delight from the perspective of my customer?" And if the answer is "No," then "Why not?"

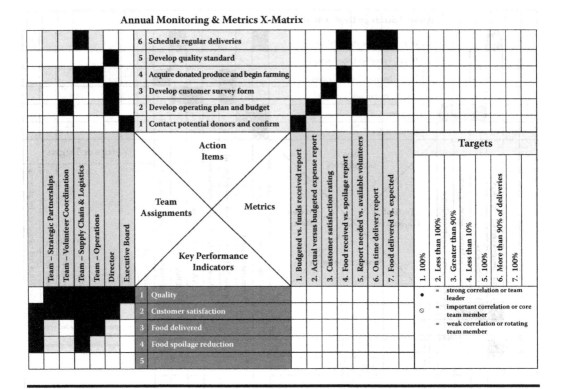

Figure 8.16

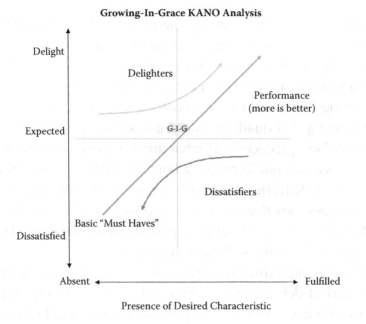

Figure 8.17

Chapter 9

Conclusion

> Well, all good things must come to an end—even those things
> some may deem delightful.

In this book I have tried to show that in pursuing the delivery of delightful service, organizations can benefit not just their customers, but themselves as well, that providing delightful service is not just "a nice thing to do." It is a strategic objective with long-term benefits all on its own, for both organizations and individuals. And how do you arrive at this destination? By embracing the Kano philosophy and by using the Lean, quality, and Six Sigma tools demonstrated in this book as well as in others, to implement the philosophy at work and in your home life. It must be pointed out, though, that exceeding customer expectations does not simply mean giving "more" or "extra." There must be value for the customer in the "additional" service or product provided at the time that it is presented. Otherwise an organization is generating the waste of overprocessing. This can be a fine line of distinction at times, but it is essential for the successful organization to know where that line resides for each of its relationships. I have also tried to stress the importance of both monitoring and metrics, and aligning both to organizational goals.

Although this book is meant to provide a step-by-step guide for how to implement the Kano philosophy in your organization or your life, don't be bound by its contents. There are many more tools and strategies that can be used to deploy Kano successfully in your organization. *The Quality Toolbox* by Nancy R. Tague is but one excellent resource. This book is meant to be a starting point on a journey to be uniquely interpreted by each organization and individual. The most important thing that I would like to leave you with is a way of thinking about your interactions with external and internal customers as well as other individuals. I hope that I have shown you the benefits of and successfully encouraged you to pursue delivering delightful service in your day-to-day interactions as well as in your long-term strategic planning. I further hope that the tools I have shared with you in this work will be of use to you in that endeavor.

To conclude, I repeat the challenge that I issued at the beginning of this book—constantly ask yourself the questions—where is the delight in this activity? Am I providing the best possible value to the people that I interact with on a daily basis, in relationships both business and personal? If the answer is no, then why not?

Thank you for your time and kind attention in reading this book.

Bibliography

Bautista Smith, J. (2012). *Auditing Beyond Compliance.* Milwaukee, WI: Quality Press.

Kano, N., Seraku, N., Takahashi, F., and Tshuji, S. (1996). *Attractive Quality and Must-Be Quality. Best on Quality, IAQ Book Series* Vol. 7, Milwaukee, WI: ASQC Quality Press, pp. 165–186.

Kobayashi, I. 1995. *20 Keys to Workplace Improvement*, revised edition. Trans. Bruce Talbot. Portland, OR: Productivity Press.

Munro, R.A., Maio, M.J., Nawaz, M.B., Ramu, G., and Zrymiak, D.J. (2008). *The Certified Six Sigma Green Belt Handbook.* Milwaukee, WI: ASQ.

Okes, D. (2013). *Performance Metrics: The Levers for Process Management.* Milwaukee, WI: ASQ Quality Press.

Petruska, B. (2012). *Gemba Walks for Service Excellence.* Boca Raton, FL: CRC Press.

Ries, E. (2011). *The Lean Startup: How Today's Entrepreneurs Use Continuous Innovation to Create Radically Successful Businesses.* New York: Crown.

Sayer, N.J. and Williams, B. (2007). *Lean for Dummies.* Hoboken, NJ: John Wiley & Sons.

Tague, N.R. (2005). *The Quality Toolbox.* Milwaukee, WI: Quality Press.

Vanzant-Stern, T. (2011). *Lean Six Sigma Practical Bodies of Knowledge.* Palo Alto, CA: Fultus.

Index